Outline of

Genetic Epidemiology

Newton E. Morton

32 figures and 56 tables, 1982

S. Karger · Basel · München · Paris · London · New York · Sydney

Newton Ennis Morton, Ph.D.
Director, Population Genetics Laboratory
Professor, School of Public Health
University of Hawaii
Honolulu, Hawaii 96822

National Library of Medicine, Cataloging in Publication
    Morton, Newton E.
    Outline of genetic epidemiology/Newton E. Morton.—Basel; New York:
    Karger, 1982.
    1. Epidemiologic Methods 2. Genetics, Medical 3. Hereditary Diseases I.
    Title QZ 50 M891o
    ISBN 3-8055-2269-X

Outline of Genetic Epidemiology

Contents

Contents

Contents vii

Preface

Mathematical analysis is an indispensable tool for research in genetic epidemiology, but it is an obstacle to communication with other scientists and students. None of the substantive issues in genetic epidemiology is clarified by mathematical rigor or computational elegance, and preoccupation with these technical details is boring and confusing to many people for whom the principles of genetic epidemiology are interesting and useful.

This book is a nonmathematical survey for students of epidemiology, genetics, and other biomedical sciences. Emphasis is on disease-related traits, but the principles apply to behavioral, anthropometric, and other characters implicitly related to disease. I hope that questions about the content of genetic epidemiology are answered herein, and that the reader will be satisfied that this new field is promising and not unduly controversial: if so, perhaps the design and analysis of some research programs will be altered. I shall be delighted if this book transmits the excitement of pitting simple theories against complex realities, and sometimes winning.

C. S. Chung, C. W. Cotterman, R. C. Elston, P. A. Jacobs, D. C. Rao, T. Reich, and G. G. Rhoads read parts of the manuscript. I am grateful for their helpful suggestions.

Newton E. Morton

1. Introduction

The concept of host factors in disease is as old as medicine, but systematic study began only when public health measures had reduced the impact of infectious disease in industrial societies. Neel and Schull (1954) were apparently the first to recognize a new discipline of "epidemiological genetics" concerned with the interaction of heredity and disease. During the next decade Blumberg edited *Genetic Polymorphisms and Geographic Variations in Disease* and Neel, Shaw, and Schull produced *Genetics and the Epidemiology of Chronic Diseases*, in which Thomas Francis remarked:

> "So when the human geneticist turns to disease and disorder in the population as his basis of genetic analysis, he is promptly in epidemiology. And when he asserts the concept of multiple factors to produce an effect, he is in full cry epidemiologically. Conversely, where the epidemiologist seeks explanation for familial or other group aggregations of health or disease, he is immediately involved in genetic problems."

By 1967 when Morton, Chung, and Mi discussed problems in genetic epidemiology, there was general agreement that synthesis of goals and methods from epidemiology and genetics was inevitable and desirable. We preferred the term genetic epidemiology because determinants of familial aggregation may be purely environmental, whereas epidemiological genetics suggests prejudice against environmental hypotheses. A formal definition is—

> *genetic epidemiology*: A science that deals with etiology, distribution, and control of disease in groups of relatives and with inherited causes of disease in populations.

Inherited, as here used in a broad sense, includes both biological and cultural inheritance. The set of relatives may be as close as twins or as extended as an ethnic group.

1.1 Genetic Epidemiology Arose from Population Genetics

The scientific study of heredity, which began less than a century ago, gradually developed into three major disciplines. Biochemical genetics is concerned with the physiochemical reactions by which genetic determinants are replicated and produce their effects. Cytogenetics deals with the chromosomes that carry these determinants. Population genetics treats the mathematical properties of genetic transmission in families and populations.

When the object of population genetics is to infer the structure of the genetic material, it is called formal genetics. If the trait under study is continuous instead of discrete, and if the aim is to improve domestic plants and animals by artificial selection, we speak of biometrical (or quantitative) genetics. If the primary goal is to understand changes in gene frequency, we have evolutionary genetics, which is often divided into theoretical and experimental branches, as if prediction and observation could fruitfully be separated. Led by the brilliant trio of Wright, Fisher, and Haldane, biometrical and evolutionary genetics reached essentially their present state by midcentury.

1.2 Its Branches Include Recurrence Risks and Behavior Genetics

Genetic epidemiology is the youngest and most rapidly developing branch of population genetics. In conjunction with formal genetics it has given rise to risk analysis, the essential element of genetic counseling. Applied to psychological traits, it is called behavior genetics, but the principles and methods are the same as for physical disease. When distorted for political ends, it becomes vigilante genetics. This is indicated in figure 1.2.1 by a dotted line, the symbol for illegitimacy.

1.3 Cultural Inheritance Introduces Complications Unfamiliar to Geneticists

Genetic epidemiology is not merely a branch of population genetics. Its other parent is epidemiology, which is largely concerned with environmental causes of disease. Whereas genetics considers the environment as noise and heredity as the signal, epidemiologists tend toward the opposite simplification. Only genetic epidemiology has (or should have) neither an environmentalist nor hereditarian bias.

The geneticist, although no salesman for heritability, is so accustomed to randomizing or otherwise controlling the environment that he is likely to founder if presented with the problem of inferring causation when the environment cannot be manipulated. The persistence of leprosy in an Acadian isolate does not prove genetic susceptibility, since close

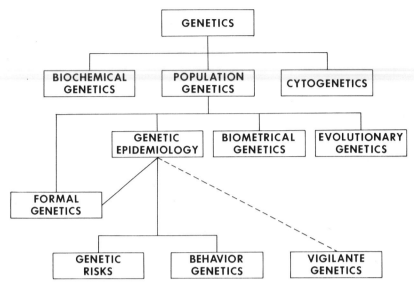

Fig. 1.2.1: The main subdivisions of genetics.

and prolonged contact is necessary for contagion of *Mycobacterium leprae*. The virtual limitation of the neurological disease kuru to the Fore tribe of New Guinea does not establish their unique susceptibility, since this slow virus disease is transmitted through cannibalism of relatives. Tuberculosis and congenital syphilis were considered hereditary until their infectious basis was understood. Geneticists did not contribute to the elucidation of these problems and in some cases stubbornly defended genetic hypotheses based on nothing more than the fact of family resemblance.

1.4 Epidemiologists Have Neglected Genetic Determinants of Disease

So long as human genetics was concerned with rare traits of simple inheritance while epidemiology dealt with common nongenetic diseases, there was little contact between the disciplines. Epidemiology could afford to ignore the elegant methods geneticists had developed to determine frequencies and familial patterns of rare diseases, free of bias due to incomplete ascertainment. Once attention shifted to diseases of complex etiology and regional frequency differences, epidemiology was as much at a loss as genetics. MacMahon and Pugh (1970) wrote:

"... it will usually be very helpful to the epidemiologist to know whether the familial occurrence of a particular disease is likely to have a genetic or an environmental explanation. If the former, then he may choose to disregard the familial pattern in forming his own hypotheses, regarding it as evidence of variation in genetic susceptibility to whatever environmental factors are involved in the etiology of the disease. If, on the other hand, the familial pattern seems likely to have an environmental explanation, the tendency to recur or persist in families might be a characteristic to be considered in developing hypotheses as to what specific environmental factors may be involved."

This implies that epidemiologists are not interested in genetic variation per se, which conflicts with the definition of epidemiology as the study of determinants of disease (not just environmental determinants). Even if this arbitrary limitation is accepted, is epidemiology willing to leave to another discipline the decision whether genetic factors are likely or unlikely causes of familial recurrence? All diseases depend both on host factors and exogenous causes, although one or the other may be preponderant for a particular disease.

Synthesis of genetics and epidemiology is necessary before diseases of complex etiology can be understood and ultimately controlled.

1.5 Questions

1. *How is genetic epidemiology distinguished from biometrical genetics?*

By absence of controlled matings or selection programs and by inclusion of major loci, chromosomal aberrations, and cultural inheritance.

2. *How is genetic epidemiology distinguished from evolutionary genetics?*

By its preoccupation with contemporary, health-related problems instead of speculation about what happened in the remote past or will happen in the remote future.

3. *How is genetic epidemiology distinguished from formal genetics?*

Formal genetics does not include polygenes, cultural inheritance, or random environment, nor is it concerned with population frequencies. It may therefore be considered a subset of genetic epidemiology.

1.6 Bibliography
Historic

Blumberg BS: Genetic Polymorphisms and Geographic Variations in Disease. Grune and Stratton, New York, 1961

Burdette WJ: Methodology in Human Genetics. Holden-Day, San Francisco, 1962

Morton NE, Chung CS (ed): Genetics of Interracial Crosses in Hawaii. Karger, Basel, Switzerland, 1967

Neel JV, Schull WJ: Human Heredity. The University of Chicago Press, Chicago, 1954

Neel JV, Shaw MW, Schull WJ: Genetics and the Epidemiology of Chronic Diseases. A symposium, June 17–19, 1963, University of Michigan Medical School, Ann Arbor, Michigan. US Department of Health, Education and Welfare, Public Health Service Publication 1163, 1965

Recent

Inouye E, Nishimura H (eds): Proceedings of the Symposium on Gene-environment Interaction in Common Diseases. Japan Medical Research Foundation Publication No 2. University of Tokyo Press, Tokyo, 1977

MacMahon B, Pugh TF: Epidemiology: Principles and Methods. Little, Brown and Company, Boston, 1970

Mielke JH, Crawford MH (eds): Current Developments in Anthropological Genetics. Vol 1. Theory and Methods. Plenum Press, New York, 1980

Morton NE, Chung CS (eds): Genetic Epidemiology. Academic Press, New York, 1978

Sing CF, Skolnick M (eds): The Genetic Analysis of Common Diseases: Applications to Predictive Factors in Coronary Heart Disease. Alan R Liss, New York, 1979

Susser M: Causal Thinking in the Health Sciences: Concepts and Strategies of Epidemiology. Oxford University Press, 1973

2. The Genetic Material

An old riddle asks, "What can an elephant do that no other animal can?", and the answer is, "Make more elephants." Conformity to inherited patterns is a conspicuous attribute of life. In important respects, these patterns are similar for all organisms, and many biological processes are carried out in the same way by bacteria, molds, and men. Other patterns are characteristic of particular species or smaller groups, such as races, clones of cells, inbred lines, or pairs of identical twins, the members of which share common features of appearance, behavior, or metabolism that distinguish them from other groups in the same environment. Conformity to inherited patterns is dependent, at the ultimate biochemical level, on the remarkable precision with which the genetic material reproduces itself.

Modern genetics began with experiments on peas by Gregor Mendel, an Austrian monk. He showed that many traits are inherited as if determined by alternative, particulate factors, combinations of which give rise to characteristic genetic ratios, such as 1 : 1 and 3 : 1. When Mendel's work was published in 1866, it attracted little notice, but when it was independently rediscovered and confirmed by three other investigators in 1900, biologists all over the world immediately turned their attention to the mechanisms of heredity. The generality of Mendel's work was established throughout the plant and animal kingdoms, from viruses to man. It was demonstrated that most inherited differences depend on these mendelian factors, or genes, which are arranged in linear order on structures in the nucleus, the chromosomes. When chromosomes are lost or added, or attached to each other, when parts of them are deleted, duplicated, or rearranged, the genes they contain have their transmission altered in predictable ways.

As knowledge was gained about the formal mechanisms of inheritance, geneticists became preoccupied with the chemical nature of genes and the manner in which they control development. This inquiry culminated in an understanding of genic structure and the code for protein synthesis, together with a few tantalizing indications of how gene action is regulated. Our study of the behavior of genes in populations may well start with a brief review of the genetic material.

2.1 The Genetic Material is DNA

Life is characterized by two classes of irregular macromolecules, the nucleic acids and proteins. For a long time, the simple composition of nucleic acids and their fragility under certain physical-chemical techniques misled geneticists into believing that these molecules have a monotonous, repetitive structure incapable of carrying genetic information. During the last 30 years, however, it has been shown that genes are deoxyribonucleic acids (DNA), except in certain viruses where ribose is substituted to give a gene composed of ribonucleic acid (RNA). For all other organisms, there is a fixed sequence for transfer of genetic information,

DNA ⟶ RNA ⟶ protein, which provides a chemical definition of life.

Early evidence for this broad generalization was indirect: the amount of DNA is remarkably constant for cells with the same set of chromosomes, and the wave lengths of ultraviolet which are most effective in changing gene structure are those that are best absorbed by nucleic acid. The first clear proof that the gene is DNA came from the observation that a highly purified extract of DNA from a virulent pneumococcus strain with a carbohydrate capsule could confer on nonencapsulated, nonvirulent bacteria the inherited capacity to synthesize a capsule having the antigenic properties and virulence of the donor strain. This transforming activity is destroyed by an enzyme that specifically degrades DNA, whereas ribonuclease (which degrades RNA) and proteases (which degrade proteins) have no effect. Later the T2 virus, which is composed of a DNA core in a characteristic protein coat, was shown by labeling protein with the radioactive isotope ^{35}S to shed its coat before infecting bacteria. DNA therefore provides all the information required to synthesize protein for the progeny (fig. 2.1.1).

2.2 DNA Has a Triplet Code

As soon as the genetic significance of DNA was recognized, its structure became the central problem of molecular biology. A single strand of nucleic acid has a repeating sequence of sugar-phosphate groups, the sugar being deoxyribose in DNA and ribose in RNA. The sugar residue attaches to a nitrogen base, of which there are four in DNA: adenine (A), cytosine (C), guanine (G), and thymine (T). In RNA, uracil (U) replaces thymine. The sequence of base pairs ultimately gives all the information necessary to synthesize proteins. Since there are four possible bases for

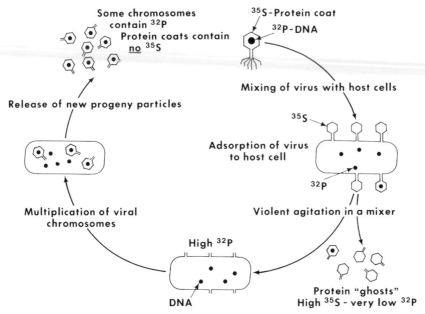

Fig. 2.1.1: Demonstration that only the DNA component carries genetic information and that the protein coat functions as a protective shell that facilitates DNA transfer.

each sugar, there are 4^n possible sequences for a nucleic acid strand containing n bases, where n is of the order of 1000 for a typical gene. This encodes a virtually infinite amount of genetic information, since 4^{1000} is larger than the number of elementary particles in the universe!

Except in some viruses, each strand of DNA is ordinarily paired by hydrogen-bonding with the bases of a complementary strand to form a ladder in which adenine is always paired with thymine (A-T or T-A) and guanine with cytosine (G-C or C-G), so that there are still only four possibilities for each base pair. In this structure the flat bases stack above each other, the two strands twisting so that the ladder becomes a double helix. Base-pairing gives the molecule two properties essential for the genetic material: it has great metabolic stability, permitting transmission through thousands of generations without change in base-pair sequence; and the strands are complementary, so that each has the information to synthesize the other. The single-stranded viruses are really no exception, since they form the complementary strand as soon as the DNA enters the host cell.

Proof that DNA replicates by strand separation and formation of complementary molecules comes from experiments in which bacteria are grown for many generations in a medium containing the heavy isotope ^{15}N until virtually all of the normal ^{14}N is replaced. Then a single replication is permitted in ^{14}N. The resultant DNA is composed of two strands, one with ^{15}N and the other with ^{14}N. A second replication in ^{14}N produces two populations of DNA, one composed of the hybrid between ^{15}N and ^{14}N strands, as in the previous generation, and the other of pure ^{14}N/^{14}N pairs. This characteristic mode of replication, in which each strand acts as a template for its complement with which it instantly pairs, is called semiconservative, since it produces a molecule with one old and one new strand.

The primary structure of both DNA and protein is linear, and therefore each amino acid in a protein must be specified by a sequence of bases in the DNA (see fig. 2.2.1). There are 20 amino acids and only four bases. Sequences of two bases could encode only $4^2 = 16$ amino acids, whereas sequences of three bases could encode $4^3 = 64$ possibilities. Evidence that the coding unit (*codon*) has exactly three bases came first from proflavine-induced mutations (see fig. 2.2.2) which consist of single base deletions and insertions. DNA with three such deletions (or insertions) in a small region, each of which is associated with loss of protein function, may produce normal protein, whereas combinations of one, two, or four such deletions (or insertions) are inactive. Clearly these combinations change the way a three-letter code is read after the first deletion or insertion, whereas exactly three deletions (or insertions) change the code only in the interstitial region.

Once it became clear the codon is a triplet (see fig. 2.2.3) several ingenious chemical methods quickly deciphered the code. Certain amino acids are completely specified by the first two bases, the third serving

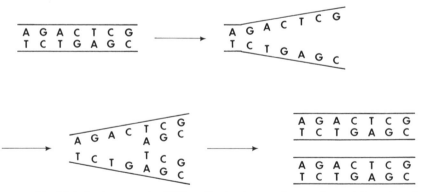

Fig. 2.2.1: Semiconservative DNA replication by strand separation and formation of complementary molecules.

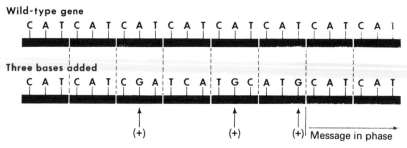

Fig. 2.2.2: Triple mutation in which three bases are added fairly close together spoils the genetic message over a short stretch of the gene but leaves the rest of the message unaffected. The same result can be achieved by deletion of three neighboring bases.

only to maintain a triplet code. Thus the four DNA triplets beginning with GT all encode the amino acid valine. The code is therefore said to be degenerate. The sequences TAA, TAG, and TGA terminate the protein chain. The code to initiate the chain is ATG. The corresponding amino acid (methionine) may be removed enzymatically, explaining why many proteins begin with other amino acids. The coexistence of initiator and terminator codons permits the genetic material to be transcribed normally, even when sequences beginning and ending with these codons are inverted. Genes in higher organisms typically include segments that do not begin with an initiator codon and are not transcribed. The function of these intervening sequences is unknown.

2.3 DNA Is Organized into Chromosomes

The DNA encoding a particular protein chain does not ordinarily occur as a separate molecule. Instead, long strings of DNA, each containing many genes, are normal features of all organisms. These multigenic DNA molecules, with the RNA and protein with which they are associated in higher plants and animals, are called *chromosomes*. Their function is to protect the genetic information, distribute it in an orderly way to the daughter cells, and regulate gene action.

The simplest chromosomes are found in viruses and bacteria, where they are often circular. There is no linear morphological differentiation and apparently no significant protein component. Only a single chromosome type is present in each cell. Bacterial chromosomes are located in nuclear areas, which, however, are unbounded by a membrane.

Chromosomes of higher organisms are considerably more complex. The DNA still runs throughout their length without apparent interrup-

1st	2nd				3rd
	T	C	A	G	
T	Phe	Ser	Tyr	Cys	T
	Phe	Ser	Tyr	Cys	C
	Leu	Ser	•	•	A
	Leu	Ser	•	Trp	G
C	Leu	Pro	His	Arg	T
	Leu	Pro	His	Arg	C
	Leu	Pro	Gln	Arg	A
	Leu	Pro	Gln	Arg	G
A	Ileu	Thr	Asn	Ser	T
	Ileu	Thr	Asn	Ser	C
	Ileu	Thr	Lys	Arg	A
	Met*	Thr	Lys	Arg	G
G	Val	Ala	Asp	Gly	T
	Val	Ala	Asp	Gly	C
	Val	Ala	Glu	Gly	A
	Val	Ala	Glu	Gly	G

Fig. 2.2.3: The genetic code. The codon marked * initiates the polypeptide chain, which is terminated by any of the three codons marked •.

tion, but there is more than one chromosome type per cell, contained within a nuclear membrane. The chromosomes are associated with substantial quantities of protein and are linearly differentiated in a pattern that serves to distinguish one chromosome type from another in the same cell. The darker-staining regions, called *chromomeres*, have marked secondary coils. Most chromosomes have a region called the *centromere*, which regulates chromosome assortment in cell division. In a few organisms this function is distributed diffusely over the whole chromosome. A specific region in certain chromosomes is the *nucleolar organizer*, where genes for ribosomal RNA are localized (see fig. 2.3.1).

Chromosomal regions adjacent to the centromere, intercalated in the rest of the chromosome, and along most of the length of the heteromorphic sex chromosome remain tightly coiled during development and replicate after the rest of the chromosome. This behavior defines *heterochromatin*, which by the usual criteria contains few active genes. In many vertebrates one of the two sex chromosomes (the inactivated X) is heterochromatic in the female. Heterochromatin tends to associate nonspecifically and appears to play a role in synapsis of homologous chromosomes.

The number of DNA molecules within a chromosome is variable among species and developmental stages. Polytene chromosomes of certain cells may have at least 1000 parallel molecules through repeated replication, but most cells have only one or a few DNA molecules per chromosome. DNA begins replication at many points along the length of a chromosome, but there is no evidence of a physical discontinuity between genes or sequences of genes.

2.4 Chromosomes Are Distributed Reductionally in Meiosis

In somatic cell division, or *mitosis*, each chromosome replicates and the products separate and are distributed equationally to the two daughter cells by spindle fibers attached to the centromere. The result is, with rare accidents, the creation of two daughter cells identical to their parent.

Meiosis, or reductional division, is much more interesting genetically, since it occurs only in the immediate precursors of the sex cells and results in transmitting to each offspring half of the parental genetic information, which is then restored to its normal value by the union of egg and sperm. If the quantity of DNA in an egg or sperm of a given species is denoted by C, the quantity in the usual somatic cell is 2C. Similarly, if the number of chromosomes in an egg or sperm is denoted by N, the somatic number is 2N. N is the *haploid* number, and the corresponding developmental stages from reduction division to fertilization is the *haplophase*. The *diploid* number is 2N, and the stages between fertilization and reduction division are the *diplophase* (see fig. 2.4.1).

Meiosis begins by DNA replication in a diploid cell, which raises the

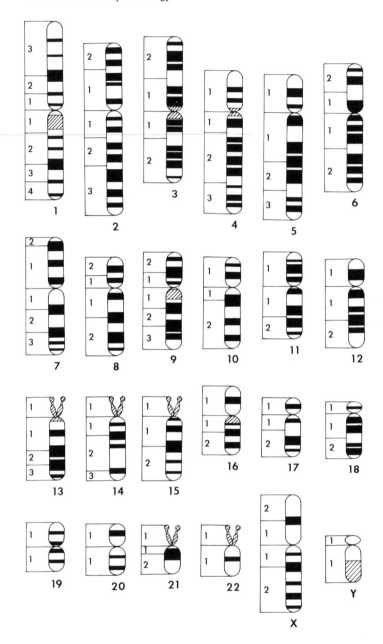

Fig. 2.3.1: Diagrammatic representation of chromosome bands as observed with the Q-, G-, and R- staining methods; centromere representative of Q-staining method only. Each chromosome is diagrammed with the short arm (p) above the centromere and the long arm (q) below it.

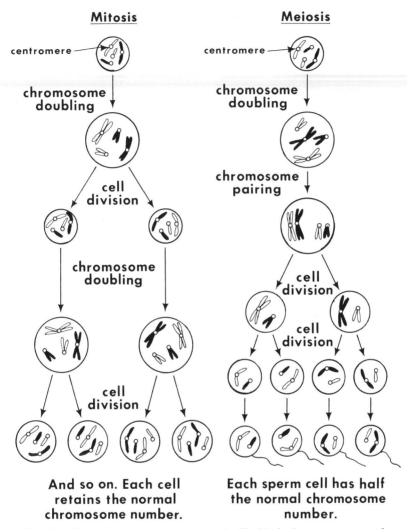

And so on. Each cell retains the normal chromosome number.

Each sperm cell has half the normal chromosome number.

Fig. 2.4.1: Comparison of mitosis and meiosis. The black chromosomes came from one parent, the white from the other.

amount to 4C. To reduce this to C in the haplophase requires two cell divisions. In the first, chromosomes of the same type (*homologues*) pair, one member of each pair having come from each parent. At this stage each chromosome is composed of two chromatids, each containing an amount C of DNA. The four chromatids in a bivalent are able to exchange corresponding parts by a process called *recombination* or *crossing-over*. The forces that bring the chromosomes into precise synapsis and determine that these exchanges be ordinarily between adjacent base pairs are not understood. As the spindle fibers attached to the two centromeres separate the homologous chromosomes, the sites of exchanges between them are apparent as contact points that become X-shaped *chiasmata* as the chromosomes pull apart.

At the end of this first meiotic division the DNA content has been reduced from 4C to 2C, and the chromosome number from 2N to N. From each pair of homologues one centromere, either maternal or paternal, has been sampled. However, the two chromatids attached to this centromere are in general different because of crossing-over. In the second cell division, the centromere divides and the chromatids are distributed to different cells. The final result is a set of four cells, each containing one chromatid from each pair of homologues. If the latter bore different genes, the chromatids will differ among themselves and from the parents. Thus if one parental chromosome had the genes A and B, and the other had a and b at corresponding loci, a single crossover between them would produce Ab and aB chromatids, in addition to the nonrecombinant AB and ab chromatids. A four-strand double crossover would produce only Ab and aB chromatids. In the absence of interstitial markers, any odd number of crossovers between A and B on two chromatids would be indistinguishable from a single crossover, whereas an even number would restore the nonrecombinant type. Ordinarily a large number of meiotic events are pooled in a sample of eggs or sperm, but special situations where two or more of the four daughter cells can be recognized have permitted a determination of some statistical properties of crossing-over among the four chromatids.

The sex of most higher organisms is normally determined by a pair of chromosomes, called the sex chromosomes, which are often distinguishable under the microscope. The remaining chromosomes are called *autosomes*. In man, for example, the female is homomorphic, with two X chromosomes, while the male is heteromorphic, with an X and a smaller Y chromosome. Meiosis provides an elegant mechanism for sex determination, the sperm which carry an X chromosome ordinarily producing females and the Y-bearing sperm, males. In some organisms, such as birds and butterflies, it is the female that is heteromorphic, while other species lack a Y chromosome or have other variations of the basic pattern.

2.5 Genes and Chromosomes Mutate

The structure of the genetic material and the processes of mitosis and meiosis provide a remarkably precise mechanism for replication and transmission. However, errors do occur, and every daughter cell is not identical to its parent. Without inherited variation, biological evolution would be impossible and the subject matter of genetics would not exist. Any inherited change in the genetic material is called a *mutation*.

The easiest mutations to understand are those that substitute one base for another. This occurs spontaneously at very low rates (less than 10^{-8} per base per cell division). Nitrous acid and other chemicals greatly increase the frequency of these reversible changes. The resultant alteration of only a single amino acid often does not seriously impair protein function, and so descendants of such mutations are commonly observed in natural populations. Other mutations involve the loss (deletion) or gain (insertion) of bases. Reverse mutation back to the original sequence is impossible for long deletions and very rare for insertions and single deletions.

Base analogues like 5-bromouracil, an analogue of thymine, cause mutations by incorporation into DNA and subsequent mistakes in base pairing. For example, 5-bromouracil can sometimes pair with guanine instead of its normal partner, adenine. Various high-energy radiations (including X-rays, ultraviolet, β-rays, and neutrons) yield mutations in proportion to the number of ionizations produced. High temperature and many chemicals are also effective. Spontaneous mutation rates are modified during evolution, being much less per unit of absolute time in organisms with a long generation span, like man, than in short-lived lower forms. Most mutations are harmful to the individual, but evolution depends on the small proportion of beneficial mutations.

Some inherited changes involve gross chromosomal aberrations. When a chromosome is broken in two places, either spontaneously or by ionizing radiation or a chemical mutagen, it may rejoin in a changed order so that the standard gene sequence ABCD becomes ACBD. Probably many such *inversions* so alter the genetic information as to be lethal in single dose. If viable, the inversion can pair with its standard homologue only if one or the other forms a loop, thus the standard sequence is ABCD; the inversion is ACBD.

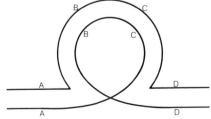

This configuration can be recognized during meiosis, or if there is somatic pairing. Inversions may also be detected by genetic tests for reduced recombination and altered sequence, and by examining mitotic chromosomes if they are linearly differentiated.

Crossing-over in an inversion loop has different consequences, according to whether the centromere is included in the inversion or not. If the inversion is *pericentric* (including the centromere), crossing-over produces deletions and duplications. Unless the region contains few genes, these chromosomes will be lethal even in single dose. If the inversion is *paracentric* (not including the centromere), crossing-over produces chromosomes with duplications and deletions and either no centromere (*acentric*) or two centromeres (*dicentric*). Acentrics have no means of being pulled to the pole in cell division and are not regularly distributed to the daughter cells. Dicentrics are pulled in opposite directions by their disjoining centromeres and are fragmented by cell cleavage, giving rise to secondary *duplications* and *deletions*. Both duplications and deletions may arise in other ways, including rare mistakes in chromosome synapsis leading to unequal crossing-over.

Inversions have a number of genetic consequences in combination with their standard homologue. They tend to increase the frequency of nondisjunction due to mechanical pairing difficulties, to reduce the apparent frequency of crossing-over in the inversion loop because some crossover products are lethal, and, if the inversion is paracentric, to form bridges and fragments when the other chromosomes are pulled to the poles in the first meiotic division. None of these consequences arise if the homologous chromosome carries the same inversion, but the frequency of crossing-over will be altered by the change in sequence. Pericentric inversions change the position of the centromere and therefore may detectably alter the ratio of the chromosome arm lengths.

Two breaks in nonhomologous chromosomes may lead to an exchange of segments (*heterologous translocation*). The translocated chromosomes, in combination with their standard homologues, produce a characteristic quadrivalent at meiosis made up of the two chromosome pairs. These may disjoin so that the two translocated chromosomes go to one pole and the two standard chromosomes to the other (*orthoploid or alternate assortment*), or the translocated and one standard chromosome may go to the same pole (*heteroploid or adjacent assortment*). Orthoploid gametes have a normal gene content, but heteroploid gametes have duplications and deletions, and are nearly always lethal. The two types of assortment are about equally frequent. If adjacent chromosomes go to the same pole, the quadrivalent opens to a ring. If alternate chromosomes assort together, the quadrivalent forms a figure-eight.

Despite the drastic effects of inversions and translocations, they are not rare in many populations and groups of related species. There are several reasons for this. Acentric and dicentric chromosomes tend to be sequestered during meiosis into nonfunctional egg and sperm, so that their effects are minimized. Gene arrangement on chromosomes is not random, and a new sequence may be beneficial. The structure of genes at the points of breakage may be altered, sometimes advantageously. Finally, both inversions and translocations restrict recombination and thereby permit the evolution of blocks of genes favorable for specific environments.

Another class of mutations comprises changes in chromosome number. Sometimes (especially when treated with colchicine), cells in mitosis complete chromosome division but fail to cleave into two daughter cells. The result is a cell with twice the diploid number of chromosomes, or 4N. Such cells are called *tetraploid*. A proportion of tetraploid cells is often found in normal tissues and newly-established tissue cultures.

In meiosis, unreduced gametes that have the diploid chromosome number are sometimes formed. If one of these unites with a normal haploid gamete, the zygote is *triploid* (3N). The union of two diploid gametes gives rise to a tetraploid zygote. Individuals that differ from normal by multiples of the haploid chromosome number (3N, 4N, and so on) are called *polyploid*. If the multiple is odd or the diploid sets of an *even-numbered polyploid* are closely related, fertility is low because of irregular assortment of chromosomes in meiosis. Such *autopolyploids* are contrasted with *allopolyploids*, where the diploid sets are so distantly related that the even-numbered polyploids segregate as diploids.

The polyploid series kN (k = 1, 2, ...) constitutes *euploidy*. Other types of variation in chromosome number, not involving multiples of the haploid set, are called *aneuploidy*. The set of aneuploids with chromosome numbers greater than 2N is called *polysomy*. The most common case is *trisomy*, in which a particular chromosome is represented three times, giving the chromosome formula 2N + 1. A number of trisomics are known in man, each associated with a particular complex of defects, or *syndrome*.

Trisomics arise from the union of a normal haploid gamete with a disomic gamete (N + 1), which resulted from *nondisjunction* of a chromosome pair, either in the first or second meiotic division. *First-division nondisjunction* occurs when homologues fail to pair or, having paired, fail to disjoin normally. In such cases the members of the pair may either assort at random to the poles, or one or both may lag at the equatorial plate and be fragmented or excluded from the daughter cells. In a fraction of cases, both homologues pass to the same pole, resulting in a disomic cell. *Second-division nondisjunction* occurs more rarely when, for reasons that are not understood, the sister chromatids fail to assort to opposite poles.

Trisomic cell lines also arise as patches of tissue descended from a cell in which nondisjunction had occurred at mitosis. If one daughter cell receives an extra chromosome which the other lacks, their descendants may form complementary *twin spots*, one $2N+1$ and the other $2N-1$.

One of the best known of the trisomic syndromes in man is *mongolism* (Down syndrome), which has an incidence at birth of about 1 in 700. Mongols are characterized by serious mental retardation, slow growth, cardiac defects, and slight physical peculiarities, including the internal epicanthic fold which suggested the name. They are extremely susceptible to infectious diseases, and many of them die during childhood. Most mongols are trisomic for chromsome 21, one of the two small, acrocentric autosomes, or G chromosomes. Others are somatic mosaics of normal and trisomy-21 cells. A small proportion are heteroploid segregants from translocation heterozygotes, containing a duplication (partial trisomy) of most of the long arm of chromosome 21.

Complementary to trisomics are monosomics $(2N-1)$, which occur when a chromosome lags at cell division and is excluded from the daughter cell. Monosomics are generally much less viable than trisomics, and no liveborn case of a monosomy syndrome has been reported for an autosome in man. However, monosomics and other aneuploids do occur in tissue culture and spontaneous abortions. Nullisomics $(2N-2)$ lack both members of a chromosome pair and are inviable except in allopolyploids. Most deletions large enough to be visible under the microscope are lethal in double dose and subvital in single dose. The corresponding duplication tends to be less severe.

2.6 Not All Genes Have Fixed Chromosomal Locations

Some DNA is found in extranuclear determinants called *plasmids*, which are transmitted independently of the chromosomes. Plasmids are rarely transmitted through sperm, which carry little cytoplasm. An individual, therefore, ordinarily inherits such organelles from his mother. Once lost, the distinctive DNA cannot be synthesized by the chromosomes, showing that plasmids are self-replicating systems. Some of them, such as kappa particles in Paramecium, the factor for CO_2 sensitivity in Drosophila, and the sex-ratio spirochaete in Drosophila, can be transmitted by infection as well as through the egg, and are therefore considered to be intracellular parasites. In the mouse, a virus causing mammary cancer is transmitted only through the mother, but foster-nursing experiments have shown that it is carried in the milk, rather than the cytoplasm.

A number of genetic factors, called *episomes*, can exist either in the cytoplasm or integrated into a chromosome. Many viruses have this property, being able to lyse the host only when free in the cytoplasm. Lysogenic cells carry the factor in its integrated state on the chromosome. After ultraviolet radiation, or spontaneously at low frequency, the virus may leave its chromosomal site and begin a lytic cycle in the cytoplasm. Another episome, the F factor, is responsible for sexuality in bacteria. Some episomes have only one possible site on a chromosome, but others can occupy one of several sites. Integrated episomes may partially suppress the action of adjacent genes.

Plasmids and episomes are interesting exceptions to the principle that most of the genetic information in a cell is carried by the chromosomes as linear arrays of genes incapable of changing their order except by inversion or translocation.

2.7 Bibliography

Mange AP, Mange EJ: Genetics: Human Aspects. Saunders College, Philadelphia, 1980

Stanbury JB, Wyngaarden JB, Fredrickson DS (eds): The Metabolic Basis of Inherited Disease. Fourth edition. McGraw-Hill, New York, 1978

Vogel F, Motulsky AG: Human Genetics: Problems and Approaches. Springer-Verlag, Berlin, 1979

3. Genes in Individuals

Information about etiology and distribution of inherited disease increases with complexity of data, from individuals to pairs of relatives to pedigrees. Here we consider basic concepts applicable to individuals. The fundamental distinction is between *genotype* (an inferred set of genes) and *phenotype* (an observed effect of these genes). Symbolically,

$$p = g + e$$

where p is the phenotype, g is genotype, and e is environment. If *e* is so nearly constant that every genotype corresponds to only one phenotype, we say that there is *complete penetrance*. A unit of DNA which produces a characteristic macromolecule (RNA or protein) is called a *locus*. The set of loci in a gamete or individual is the complete genotype, or *genome*.

3.1 Genetic Systems

The molecular structure and organization of genetic loci is poorly understood in higher organisms, partly because of the limited power of recombination tests to discriminate closely linked loci from alleles. We are therefore led to inquire what kind of genetic unit a complex system may be said to represent. "Deletions" unsupported by cytological evidence and "position effects" not demonstrated by recombination are ambiguous. The most that can be said is that the "alleles" of such a system may interact, rarely cross over, and their phenotypic effects are nonrandomly associated. Often a system is known or suspected to include two or more loci, arising either through duplication and subsequent divergence or by selection for tight linkage of structurally different but interacting loci.

We are, therefore, led to avoid the term "locus," except where its RNA or polypeptide product is well characterized, and to consider instead the more general concept of *system*, defined as *the unit of closely linked genetic information whose phenotypic factors are nonrandomly associated in panmictic populations of higher organisms*. Panmixia (random mating) signifies that alleles unite entirely at random; this concept is considered in detail in 3.6. The alternative genetic forms of a system may without ambiguity be called *alleles*. If the system is known to include two or more loci its alternative forms are often called *haplotypes*, and alleles are restricted to products of single loci. Noninteractive linked loci are excluded from a system by the requirement that the population in which the allele frequencies are determined must be panmictic and therefore indefinitely large: it is essential that virtual panmixia have persisted for a sufficient number of generations that linkage disequilibrium be insignificant. Classification of loci into systems evolves with increasing knowledge. The term *gene* usually means an allele, but sometimes designates a locus. A *marker* or *factor* is a property of an allele or a set of alleles.

3.2 Human Gene Nomenclature

Recently guidelines have been formulated for an international system of human gene nomenclature. The main provisions are:
1. Loci (and systems) are designated by capitalized letters or capitalized letters and Arabic numerals. Up to four characters is strongly recommended, and every effort should be expended to maintain this upper limit of characters. These symbols are either underlined or italicized.
2. The initial character should desirably be the first Latin letter of the gene product.
3. All characters should be written on the same line, without subscripts or superscripts.
4. Greek letters prefixing a locus name should be replaced by the Latin equivalent and placed at the end of the symbol. Thus α-fucosidase is the product of the *FUCA* locus.
5. When gene products of similar function are coded by different loci, they are designated by Arabic numerals as suffixes. For example, *PGM1*, *PGM2*, and *PGM3* are three loci for phosphoglucomutase activity.
6. The final character in a locus symbol may be used to specify a unique property within a class of loci. Thus *PKL* determines a liver form of pyruvate kinase.
7. Allele designations are written on the same line as locus symbols, separated by an asterisk.
8. The allele symbol should be limited if possible to four characters. Only capital letters and Arabic numerals in any order may be used, italicized or underlined.
9. Genotype symbols are the concatenation of allele symbols, separated

by a slash. When there is no ambiguity the locus and asterisk symbols may be omitted, as 1/2 for *ADA**1/*ADA**2.

10. Phenotype symbols have the same characters as the locus and allele symbols. They should not be underlined or italicized, and all characters should be written on the same line. An asterisk is replaced by a space. For example, *ADA* 1-2 is the phenotype corresponding to the *ADA**1/*ADA**2 genotype.

Special conventions apply to loci defined clinically, structurally, antigenically, or by specific response to a virus, and to indicate quantitative effects, as indicated in the following examples: (a) *HBB**6V designates the allele responsible for sickle cell hemoglobin, which has the amino acid valine (V) substituted in the sixth codon; (b) *E1**Q0 is a "silent" allele (with no detectable enzyme activity) at the *E1* locus for cholinesterase activity; (c) for loci defined by clinical effects, initial letters N, D, and R designate normal, dominant, and recessive alleles in terms of their pathological effects (not biochemical or other properties); and (d) *EVS1* is a locus for ECHO 11 virus sensitivity. For historical reasons, these rules have certain recognized exceptions, but they provide a reasonable standard.

Practical problems in applying this notation are most severe for the disease-related loci of most interest to genetic epidemiology. Evidence that the same disease (like achromatopsia or xeroderma pigmentosum) depends on different loci in different families is often nonexistent, or based on inconclusive clinical impression. Such distinctions can be made rigorously in terms of substantially different activity or structure of the gene product, complementation in heterozygotes or in vitro, and linkage to different markers. For example, two coagulation disorders on the X-chromosome are known to be nonallelic because hemophilia (*HEMA*) is a deficiency of factor VIII, while Christmas disease (*HEMB*) involves factor IX; a mixture of plasma from patients with the two diseases has normal coagulation time (a positive complementation test); and linkage analysis shows different locations with respect to the sex-linked system for color-blindness (*CB*). Lacking such evidence, there would have been no reasonable alternative to lumping the two disorders as if they belonged to a single system (say *HEM*). The argument for lumping alleles is weaker, and suggestive clinical evidence that the disease differs substantially among, but not within, families would be sufficient. For example, dominant genes for elliptocytosis are grouped into two loci, *EL1* and *EL2*, by linkage of the former to *RH*. It may turn out that *EL2* will be split into two or more loci. Meanwhile, different clinical forms of the RH-unlinked types could be tentatively given allele designations *EL2**D1, *EL2**D2, and so on. Conservation in assigning locus designations avoids a large class of invalid symbols that cannot, without ambiguity, be reused for other loci, which is especially inconvenient if a symbol like *A2* is invalidated, so that loci of similar activity must form a series *A1*, *A3*, and so on. With the 26 letters and 10 Arabic numbers $(0, \ldots 9)$, under the above rules there are

		26	single character symbols
26×36	=	936	double character symbols
26×36^2	=	33,696	triple character symbols
26×36^3	=	1,213,056	quadruple character symbols
Total	=	1,247,714	locus and system symbols

The number would be much greater if other symbols (., −, and so on) were permitted. For alleles that may begin with an Arabic numeral, there are

$$36 + 36^2 + 36^3 + 36^4 = 1,727,604 \text{ symbols}$$

Thus, four characters code for more than the number of recognizable alleles (excluding multiple mutations and unequal crossovers) so there is no need to exceed four characters for either loci or alleles. Symbols with three characters lend themselves to the addition of a suffix when splitting becomes necessary, whereas symbols with four characters must be abbreviated if addition of a suffix is not to exceed the recommended limit. Short symbols are more easily remembered and communicated, whereas long symbols that convey more detail are impenetrable, especially for genotypes (eg, *ABCDEF*123456/ABCDEF*78910*). Particularly for antigenic systems not all factors in a system may be observed, and then ad hoc allele symbols may be required.

The next step is for an international working group to prepare a catalog of systems, loci, and alleles, such as already exists (with somewhat different notation) for Drosophila and the mouse. Mathematical genetics follows the older convention in which loci are distinguished by subscripts and alleles by superscripts; alleles need not be italicized, and there is no slash between alleles in genotype symbols. This is acceptable when the loci and alleles are general concepts, but human gene nomenclature should be followed to designate particular loci and alleles.

3.3 Factor-union Phenotype Systems

Without loss of generality we can distinguish alleles by binary attributes called factors, the presence of a factor being indicated by 1 and its absence by 0. For example, the *ABO* system (table 3.3.1) may be represented by three factors for the *A1*, *A*, and *B* antigenic structures. These three factors define four alleles determining six phenotypes. The set of genotypes with the same phenotype is called a *phenoset*.

The *ABO* system is an illustration of a *factor-union system*, in which the presence of each factor is dominant to its absence, and penetrance is complete. In the absence of mutation, this is exactly represented by the union rule of Boolean algebra: $1 \cup 1 = 1 \cup 0 = 0 \cup 1 = 1$ implies that a child has a factor if either parental gamete had it, while $0 \cup 0 = 0$ implies that a child lacks a factor unless at least one parental gamete had it. This coincidence with Boolean algebra gives factor-union systems the abstraction required to represent many different genetic systems and to communicate this information. Cotterman has derived useful properties of factor-union systems, including the number of genotypes in a given system, the composition of each phenoset, and the rules for parentage exclusion.

Only for an antigenic system is there a simple correspondence between a factor and a reagent. More generally a factor may be an electrophoretic band or other property characteristic of a homozygote. For example, the haptoglobin (*HP*) system with phenotypes 1-1, 1-2, and 2-2 is represented by binary vectors 10 and 01 for alleles *HP1* and *HP2*, respectively. Here the heterozygote phenotype is $10 \cup 01 = 11$, which is uniquely defined without considering bands peculiar to the heterozygote.

It is convenient to have a terminology for pairs of factors (a, b) within a system. There are seven different relations, depending on the absence (or great rarity) of one or more of the four possible alleles (11, 10, 01, and 00).

Case 1. Permuted factors. All four alleles and their corresponding phenotypes are present.

Example: factors C and D of the RH system

Table 3.3.1: The *ABO* factor-union system

Alleles	Phenotypes	Binary system	Antigenic factors			Phenoset
			A_1	A	B	
0	0	000	0	0	0	*0/0*
A1	A1	110	+	+	0	*A1/A1, A1/A2, A1/0*
A2	A2	010	0	+	0	*A2/A2, A2/0*
B	B	001	0	0	+	*B/B, B/0*
—	A1B	111	+	+	+	*A1/B*
—	A2B	011	0	+	+	*A2/B*

Case 2. Segregant factors. The 11 allele is missing, but all four phenotypes occur.

Example: *A* and *B* of the *ABO* system

Case 3. Codominant factors. The 00 allele and phenotype are missing.

Example: factors C and c of the *RH* system in Negroes.

Case 4. Complementary factors. Both the 11 and 00 alleles are missing, together with the 00 phenotype.

Example: factors C and c of the *RH* system in non-Negroes.

Case 5. Subtypic factors: *a* subtype of *b* (a $\subset$ b). The 10 allele and phenotype are missing. The relation a $\subset$ b is read "*a* is a subset of *b*".

Example: Factors A_1 and *A* of the *ABO* system.

Case 6. Subtypic factors: *b* subtype of *a* (b $\subset$ a). The 01 allele and phenotype are missing.

Example: factors *A* and A_1 of the *ABO* system.

Case 7. Identical factors. Both the 10 and 01 alleles and phenotypes are missing, so that *a* and *b* are on the evidence indistinguishable.

Example: two different anti-B reagents in the *ABO* system.

The examples for cases 3 and 4 illustrate that relations among factors depend on population allele frequencies.

Since these relations for factor pairs are defined on alleles, they cannot be applied to two systems that interact phenotypically. (A set of interactive systems is called a *hypersystem*.) By definition all nonallelic factor pairs are permuted. Thus the Le(a) factor on red cells and the Se factor do not ordinarily coexist phenotypically after infancy, but all four possible pairs of Le and Se factors occur in the gametes.

3.4 More General Phenotype Systems

The one-to-one correspondence between alleles and their products (one gene, one polypeptide) is reflected by factor-union systems. However, once the phenotype becomes remote from the immediate gene product, two complications may arise:

1. The phenotype system may no longer be factor-union but still be *regular*: ie, under specified conditions of classification and for a specified population, each genotype corresponds to one and only one phenotype. Regular phenotype systems that are not factor-union manifest one or both of the properties called *metataxy* and *parataxy*. Two alleles are *metatactic* if the two homozygotes are indistinguishable, but the heterozygote is distinct. Two alleles are *paratactic* if all three genotypes are phenotypically indistinguishable.

A regular phenotype system can be represented diagrammatically by a *phenogram* in which homozygotes are symbolized by filled circles (●), heterozygotes by open circles (○), and a line connects indistinguishable phenotypes. *Isomorphic* systems have the same phenogram. The four 2-allele phenograms are shown in figure 3.4.1. Metataxy may be illustrated by the use of *AB* red blood cells to classify serum phenotypes: the serum of *AB* persons does not react, whereas all other sera cause agglutination. Parataxy may be illustrated by any undiscovered system, awaiting the right technique and sample, or more interestingly by a third allele that gives distinguishable heterozygotes. Itano has proposed such a phenotype system for the β chain of hemoglobin to account for the observation that *AS* heterozygotes seem to have at least two different ratios of normal to abnormal hemoglobin. This bimodality is not evident in *AA* homozygotes. Figure 3.4.2 is the phenogram corresponding to the hypothesis of two normal isoalleles, *A* and *A'*. Two paratactic alleles distinguishable in heterozygotes with a third allele are called *isoalleles*.

Of the four 2-allele phenograms, two are factor-union systems. There are 52 three-allele phenograms, of which 9 are factor-union systems. The number of phenograms with four alleles is 5525, of which 81 are factor-union systems. The number of phenograms increases at a staggering rate with the number of alleles. Very few of the regular, nonfactor-union systems have been observed.

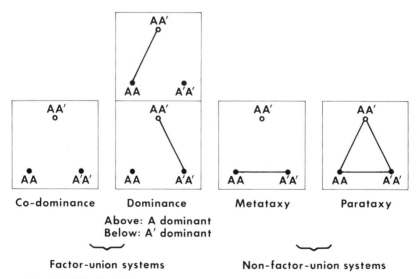

Fig. 3.4.1: The four 2-allele phenograms.

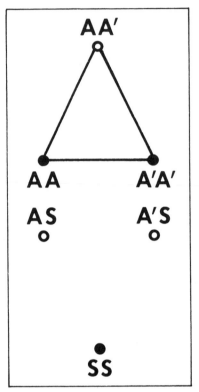

Fig. 3.4.2: The 3-allele phenogram for two isoalleles, *A* and *A'*, which are distinguishable in heterozygotes with a third allele, *S*.

2. A much more important deviation from factor-union systems is irregularity (*incomplete penetrance*, see fig. 3.4.3): ie, under specified conditions of classification and for a specified population, at least one genotype corresponds to two or more phenotypes. We could represent this by a phenogram with broken lines between overlapping phenotypes. Penetrance is often dependent on age, sex, and other factors, some of which are manipulable.

We see that the three possible deviations from a factor-union system are irregularity (incomplete penetrance), metataxy, and parataxy. All of these complications may be regarded as errors of measurement, the price the geneticist pays for having failed to characterize accurately the immediate gene product. Thus with improved techniques, all phenotype systems should in principle become factor-union systems.

One of the goals of genetics is to devise conditions under which, without hazard to the individual, his genotype can be recognized, usually through examination of blood, other body fluids, or somatic cells, with or without challenge by a diagnostic drug. Until such techniques are completely successful, genotype probabilities must be used.

3.5 Gene Frequency and Related Concepts

The geneticist is faced with astronomical variety. Even with only 2 alleles and 1000 systems, the number of possible diploid genotypes is 3^{1000}. Since it is impossible to draw precise conclusions about many systems considered simultaneously, it is often convenient to fix attention on a single system, ignoring the others. The situation can be made still simpler by considering a population not during the diplophase, with its bewildering variety of genotypes, but during the haplophase as a collection of gametes.

Given a set of N gametes (called a *gene pool*), of which $R_1, R_2, \ldots R_m$ carry allele $A^1, A^2, \ldots A^m$, respectively, we shall say that $q_i = R_i/N$ is the *gene*

A Regular System

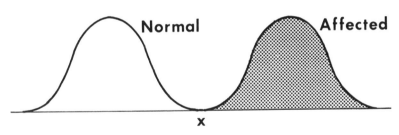

An Irregular System

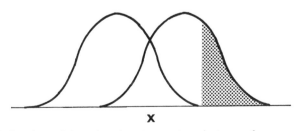

Fig. 3.4.3: Regular and irregular phenotype systems in terms of a gene product X with dominance.

frequency of A^i. Note that $\Sigma_i R_i = N$, and therefore $0 \le q_i \le 1$ and $\Sigma_i q_i = 1$.

We shall sometimes apply this definition to specified values of N. For example, consider the case $N = 2$. A homozygote $A^i A^i$ has gene frequencies $q_i = 1$, $q_j = 0$ $(i \ne j)$. A heterozygote $A^i A^j$ has gene frequencies $q_i = q_j = 1/2$, $q_k = 0$ $(k \ne i, j)$. Often, however, the value of N is indefinitely large.

Two gene pools are different if their gene frequencies are different. At this point we note that in general the gene pool of one generation is different from the next, and the gene pool of a group is different from the gene pool of a population. Population genetics is the study of relations among gene pools of size $N = 1, 2, \ldots \infty$, which differ in parentage, and therefore often in time or space.

Since individuals are sets of genes, groups are sets of individuals, and populations are sets of groups, we must consider how to pass from one gene pool to another. If the frequency of the ith allele in the kth group is q_{ki}, and if the kth group makes up a fraction f_k of the population, then the overall gene frequency is $q_i = \Sigma_k f_k q_{ik}$. (A mathematician would say that f_k is the probability distribution of the random variable q_i.)

To illustrate these ideas, consider a population consisting of two villages, the first with 100 inhabitants and the second with 200. The reagents anti-M and anti-N give the following frequencies:

Phenotype	Vector	Village 1	Village 2	Total
M	10	9	72	81
MN	11	42	96	138
N	01	49	32	81
Total		100	200	300

Here $f_1 = 1/3, f_2 = 2/3$. Since the M gene frequency is 1 in M phenotypes and $1/2$ in MN phenotypes, we have

$$q_{11} = \frac{9 + 42/2}{100} = .3 \qquad \text{for village 1}$$

and

$$q_{12} = \frac{72 + 96/2}{200} = .6 \qquad \text{for village 2}$$

In the population,

$$q_1 = \frac{1}{3}(.3) + \frac{2}{3}(.6) = .5$$

$$= \frac{81 + 138/2}{300}$$

We now leave the mathematical concepts associated with gene frequencies to consider some purely genetic notions. In our study of population genetics we shall find that rare genes differ from common ones in many ways: in their pathological effects, the way in which information is sampled, methods for estimating genetic parameters, variability in frequency among populations, effects of inbreeding, and the systematic forces that determine gene frequencies. It is convenient to have a terminology that distinguishes the main frequency classes, without implying other differences that should appear in hypotheses rather than definitions.

Alleles are called *idiomorphs, polymorphs,* or *monomorphs,* if their frequencies in a given population are less than .01, between .01 and .99, or greater than .99, respectively (table 3.5.1). To avoid the problem that many DNA sequences in a degenerate code can give the same polypeptide, so that there may not be polymorphs or monomorphs at the DNA level, we agree to distinguish between alleles only if their products are different.

A system is called a polymorphic system (*polymorphism*) if it has one or more polymorphic alleles, and a monomorphic system (*monomorphism*) if it has a monomorphic allele. Idiomorphs occur in both systems. Clearly these definitions are based on frequencies alone, not on gene effects. Idiomorphs are sometimes called pathological or private alleles, depending on whether or not they have any conspicuous effect on fitness. Monomorphs are sometimes called public or wild-type alleles. In many systems there are two or more polymorphs, none of which has any special claim to being the "wild-type", a term therefore best avoided.

The Kell (K) system provides a good example of these concepts. The Cellano allele *K*2* is common in all populations, being monomorphic in Orientals and polymorphic in other groups. The Kell allele *K*1* is polymorphic in Caucasians, but idiomorphic in other races. The Sutter allele *K*4* is polymorphic in Negroes, but idiomorphic elsewhere. The remaining alleles are all idiomorphs.

3.6 Panmixia

A diplophase generation begins when gametes from a gene pool are combined pairwise into zygotes according to some rule; these zygotes experience migration, mutation, and differential mortality and fertility; and the generation terminates with the haplophase gene pool of the next generation. By *panmixia* we mean that gametes unite entirely at random,

Table 3.5.1: Allelic genes of the Kell blood group system

| Genes | | Binary notation | Antigenic factors | | | | | | | Approximate gene frequency | | |
Name	Synonym		K1 (K)	K2 (k)	K3 (Kpᵃ)	K4 (Kpᵇ)	K5 (Ku)	K6 (Jsᵃ)	K7 (Jsᵇ)	Caucasians	Negroes	Mongoloids
K*1	K	1001101	+	0	0	+	+	0	+	0.045	0.005	0
K*2	k	0101101	0	+	0	+	+	0	+	0.935	0.851	0.99
K*3	kᵖ	0110101	0	+	+	0	+	0	+	0.01	?	?
K*4	kˢ	0101110	0	+	0	+	+	+	0	0	0.144	0
K*5	kᵒ	0000000	0	0	0	0	0	0	0	0.01	?	?

K1 = Kell; K2 = Cellano; K3 = Penney; K4 = Rautenberg; K5 = Peltz; K6 = Sutter;
K7 = Matthews

with no restriction due to finite population size, inbreeding, or assortative mating, and are enumerated before differential selection has acted. We shall say that a diploid population is panmictic if and only if the probability that a zygote receive A^i from the sperm and A^j from the egg is $q_i q_j$, and this is true for all i and j. It follows that the frequency of the genotype pair $A^i A^j$, $A^k A^l$ is proportional to $q_i q_j q_k q_l$, the proportionality constant being an integer (1, 2, 4, or 8) depending on whether reciprocal genotypes and pairs are pooled.

Panmixia is sometimes called random mating, an imprecise term also used in the sense of random pairing of zygotes. In small populations with sexes separate this is not the same as panmixia. For example, consider a population of one $A^1 A^2$ and one $A^2 A^2$ individual. Under random union of zygotes with sexes separate the only permissible mating is $A^1 A^2$ $\times A^2 A^2$, with $p_{11} = 0$. Under panmixia we have $p_{11} = q_1^2 = 1/16$. The same result is obtained under the mathematically equivalent scheme of random union of zygotes with self-fertilization permitted, since 1/4 of all pairs are $A^1 A^2 \times A^1 A^2$, and of the offspring of this mating 1/4 are $A^1 A^1$, so that $p_{11} = (1/4)^2 = 1/16$. As the population size increases, the effect of separate sexes becomes negligible.

The mature reader will have noted that panmixia is an idealization which does not faithfully represent sexual reproduction. Some organisms prefer self-fertilization and others avoid it, but few if any species practice random self-fertilization. Many other factors can cause real or apparent deviations from panmixia, of which inbreeding, selection, and misclassification of genotypes are the most important.

Despite these difficulties, the concept of panmixia has proved to be remarkably useful in population genetics, just as the idealizations of perfect vacuums and dimensionless points have served other sciences. Under a great variety of conditions cross-fertilizing populations are close enough to panmixia for this model to provide a good approximation to the genotype frequencies p_{ij}, or reasoning backward from phenotypes to estimate with reasonable accuracy the gene frequencies q_i.

Regardless of the initial phenotype frequencies, a single generation of panmixia by definition establishes the condition that $p_{ij} = q_i q_j = p_{ji}$. Thereafter, in the absence of factors that change the gene frequencies, a panmictic population large enough for chance fluctuations in gene frequency to be negligible must retain the same genotype frequencies. This principle, which was independently stated by Hardy and Weinberg in 1908, may be phrased more formally.

The Hardy-Weinberg law. An indefinitely large population reaches equilibrium genotype frequencies for a neutral autosomal system in a single generation of panmixia, regardless of the initial genotype frequencies. The equilibrium frequencies are

$$A^i A^i : p_{ii} = q_i^2$$

$$A^i A^j : p_{ij} + p_{ji} = 2q_i q_j \qquad (i \neq j)$$

where $q_1, q_2, \ldots q_m$ are the gene frequencies of alleles $A^1, A^2, \ldots A^m$.
One reason for the generality of this theorem is that if the gene frequencies are perturbed for any reason, equilibrium is reached at the new gene frequencies in the next generation, rather than slowly over a number of generations. Separate sexes are a slight exception representing a transient subdivision of the gene pool. Suppose that for some reason the gene frequencies are perturbed more in one sex than the other, so that q_i^p and q_i^m are the frequencies of A^i in paternal and maternal gametes, respectively. Clearly random union of egg and sperm will produce genotype frequencies $p_{ij} = q_i^p q_j^m$, which is not the same as for completely random union of gametes (panmixia). However, the gene frequencies are now $q_i = (q_i^p + q_i^m)/2$ for both sexes (in populations large enough for chance fluctuations to be negligible), and so a second generation of random union of eggs and sperm will produce panmictic genotype frequencies.

3.7 Sex-linked Genes

We have just seen that separate sexes delay Hardy-Weinberg equilibrium for only a single generation with autosomal genes. However, sex-linkage provides a subdivision of the gene pool which persists for many generations. Consider a species like man in which the male is heterogametic. The equilibrium expectations are:

Males	Females
$A^i Y$ q_i	$A^i A^i$ q_i^2
	$A^i A^j$ $2q_i q_j$

If, for any reason, the gene frequencies are different between males and females in any generation, the approach to equilibrium will be oscillatory: in each generation the gene frequency in females is only half as far from the equilibrium value as it was in the previous generation, but in the opposite direction. Males follow this oscillatory approach to equilibrium with a lag of one generation. Even with a large initial gene frequency difference, a close approximation to panmixia is reached within a few generations.

3.8 Two Systems

Contrary to the situation with a single autosomal locus, the equilibrium relation between two or more systems on the same or different chromosomes is not attained immediately, but gradually over a number of generations. For two systems with recombination rate θ, the deviation of gametic frequencies from equilibrium is reduced each generation by a fraction θ. The time to go halfway to equilibrium is

$$t = \frac{\ell n(1/2)}{\ell n(1 - \theta)} \tag{3.8.1}$$

$$\doteq .693/\theta \quad \text{for small } \theta.$$

Thus unlinked systems go halfway to equilibrium in a single generation, but with a recombination frequency of .001 the time is 693 generations. Until equilibrium is reached the genotype AB/ab (with A, B in *coupling*) has a different frequency from the genotype Ab/aB (with A, B in *repulsion*). When these frequencies are equal, we say that there is *gametic equilibrium* (also called *linkage equilibrium*).

Once equilibrium is reached, gametes follow Hardy-Weinberg proportions as if they were alleles. The two linkage phases (coupling and repulsion) are equally frequent, and the probability of any gamete, say A^iB^j, can be obtained by simple multiplication as $q_i^A q_j^B$. The frequency of the genotype $A^iA^iB^jB^j$ is $(q_i^A)^2(q_j^B)^2$, and of the genotype $A^iA^iB^jB^k$ is $(q_i^A)^2(2q_j^A q_k^B)$, while the frequency of the genotype $A^hA^iB^jB^k$ is $(2q_h\,q_i)(2q_j\,q_k)$, where $h \neq i$ and $j \neq k$.

Clearly a population that has been panmictic for many generations should show few effects of linkage on phenotype frequencies. Genetic factors in different systems should be distributed independently. However, recent migration may still be detectable, especially for closely linked genes. An example is provided by two sex-linked loci in man, *G6PD* (glucose-6-phosphate dehydrogenase) and *CB* (color blindness), which have a recombination frequency of about .01 in females. Since no recombination takes place for sex-linked loci in males, this gives $\theta = (2/3).01$ per generation. Equation 3.8.1 shows that about 100 generations are required to go half-way to gametic equilibrium. Populations that started with a chance excess of coupling or repulsion between these loci should retain about half the deviation after 2000 years. Conversely, populations that are combinations of many gene pools should approximate gametic equilibrium. The expectation is in fact realized, with a great excess of repulsion in Kurdistan Jews, of coupling in Sardinian villages, and gametic equilibrium in American Negroes.

An autosomal example of persistent gametic disequilibrium is provided by an experiment with recessive lethal genes in Drosophila. The investigators were interested in the possibility that recessive lethals might, on the average, confer some selective advantage on the heterozygote. They identified lethal chromosomes in the laboratory and then released large cultures of heterozygous lethals in the islands of Angra dos Reis off the coast of Brazil. When natural food sources were abundant the lethal frequencies decreased slowly, but in the season of famine the lethals rapidly disappeared at a rate so astonishing that the massive experiment was repeated over the next few years, with the same result. Equation 3.8.1 tells us that the test of lethal effects was vitiated by gametic disequilibrium. A few generations of random mating are not sufficient to establish independence between the recessive lethals and disadvantageous genes from the laboratory, whether on the same or different chromosomes.

The extension of these concepts to three or more loci is straightforward, but seldom useful. Gametic equilibrium is a reliable approximation except in hybrid populations or with tight linkage. However, factors within a locus and loci within a system are usually in disequilibrium. Evolutionary geneticists debate but are rarely able to determine whether this is partly due to interactive effects on fitness, or entirely due to the long time required for tightly linked genes to approach equilibrium after disturbance by chance, selection, or hybridization.

3.9 Gene Frequency Estimation

Given a sample of factor-union phenotypes from a panmictic population, intuition suggests more than one way to estimate the vector of gene frequencies. There is literally no end to the number of estimates we could plausibly make. Surely we don't want to conduct a detailed study of the advantages and disadvantages of alternative procedures for every estimation problem. Fortunately there is a simple solution. Under certain conditions, the method of *maximum likelihood* has been shown to give optimum estimates. When these conditions are met one can disregard all other estimates, since they can at best be as good as, but no better than, the maximum likelihood estimates.

Common sense suggests that when the probabilities of all alternatives are known, at least to a close approximation, and the stakes (ie, the consequences of winning or losing a bet) are not specified, it is reasonable to bet on the alternative with the highest probability. This gives the basic principle of maximum likelihood: take as the estimates values that maximize the probability of the observations. We shall not consider here the mathematical properties or numerical techniques of maximum likelihood estimation, but merely indicate that it is the method of choice for all problems of estimation and statistical tests of hypotheses in large samples. Experience has shown that there are few circumstances in which samples are so small that any other method is preferable. We are, therefore, free (and would be wise) to concentrate on biological aspects of estimation.

Estimation of gene frequencies and all other parameters depends on certain assumptions. When these assumptions are correct, the estimates provide useful information about the population. Sometimes, however, the assumptions are seriously in error, and then estimation procedures may reveal these discrepancies and perhaps correct for them. Estimation and hypothesis testing are complementary in genetic epidemiology, as in all science.

In table 3.9.1 a large sample of ABO blood groups is found to have an enormously significant deficiency of the AB phenotype. This not uncommon deviation is due in large part to weakening of the anti-A reaction in the AB genotype, leading to misclassification as group B. This technical error, which can be avoided serologically, has no effect on gene frequency estimation when phenotypes B and AB are pooled, giving a factor-union system with three phenotypes.

Many populations are so close to panmixia that phenotype frequencies (especially for a factor-union system) can give information about which genes are present in a population and estimates of their frequencies. A desirable condition is that no alleles are assumed which have not been demonstrated by family studies: unfortunately, with new factors such evidence may lag far behind population studies. Postulation of idiomorphs for random samples, not supported by family data, must be regarded with caution as a possible, but by no means the only, interpretation of the data, requiring confirmatory family studies or the discovery of factors specific for the inferred idiomorphs.

A satisfactory gene frequency model should meet four conditions:

1. It can account for all observed phenotypes except those attributed to technical errors,
2. It postulates few rare "alleles" that might be due to phenotypic misclassification,

Table 3.9.1: ABO blood groups in Switzerland

Separating B and AB

Phenotypes	Binary code	Observed number	Expected number (F = 0)	χ^2
A	10	130,201	129,105.53	9.30
B	01	23,263	22,018.21	70.37
AB	11	8,327	9,626.30	175.37
O	00	113,873	114,913.95	9.43
Total	...	275,664	......	264.47

Genes	Binary code	Frequency	Standard error
A	10	.295205	.000675
B	01	.059146	.000322
O	00	.645649	.000712

Pooling B and AB

Phenotypes	Binary code	Observed number	Expected number (F = 0)
A	01	130,201	130,201.00
B + AB	11	31,590	31,590.00
....	...	....	......
O	00	113,873	113,873.00
....	...	275,664	......

Genes	Binary code	Frequency	Standard error
A	01	.298241	.000703
B	11	.059041	.000322
O	00	.642718	.000730

3. The estimation procedure converges, which often requires that alleles
    not demonstrated in the sample be assumed absent,
4. The goodness-of-fit to Hardy-Weinberg proportions is acceptable as
    judged by the usual χ^2 formula or the likelihood ratio criterion.

When these conditions are not met, significant deviations may have
several causes, of which only the first two are likely to cause gross
discrepancies for polymorphs: (a) failure of the factor-union assumption
through misclassification or multiple systems; (b) rare alleles not in the
model; (c) inclusion of nonexistent alleles in the model; (d) nonrandom
sampling, for example of relatives of unusual phenotypes; (e) confounding of populations with different gene frequencies. Suspicion is directed
especially against rare phenotypes. There is a natural division between
idiomorphs, many alternative sets of which can account for rare phenotypes with or without typing errors, and polymorphs that cannot fail to
be recognized in a sample of reasonable size.

Estimates of gene frequencies for irregular phenotype systems require ancillary parameters, reflecting probabilities of misclassification. In
the simplest case of incomplete penetrance, a fraction f of individuals
with a certain genotype are classified as "affected," where f may depend
on age, sex, or other recognized factors. For idiomorphs it is often not
practical to determine phenotype frequencies from a random sample of
the population, and then special methods are required to estimate frequencies from incompletely ascertained cases (see section 8.1).

3.10 Gene Frequencies in Finite Populations

As a mathematical concept, the gene pool is a sample space with its
associated gene frequencies. As a biological concept, the gene pool is
often defined rather vaguely on a finite set of interrelated persons. This
creates some philosophical problems.

If all individuals in a community are typed for a codominant system,
in what sense is there a sampling error? Why do some anthropologists
pride themselves on avoiding relatives in an otherwise random sample?
If it is not meaningful to ask whether the gene frequencies of two individuals are "significantly" different, how can it be valid to test the significance of a gene frequency difference between two communities or generations? A solution to these troubling problems should recognize both the
convenience of maximum likelihood theory and its limitation to independent trials, and therefore to an infinite population.

If we suppose that individuals in a sample are drawn at random
from the infinitely large number of children that the preceding gener-

ation might have had, then the gene pool is infinite and with gene frequencies exactly the same as the finite preceding generation. This logic allows us to use maximum likelihood theory in finite populations with good conscience, if sampling is random (or exhaustive).

A more difficult problem arises when sampling is not random by individuals but by families, as in multigeneration data or when some families in the population are more accessible or cooperative than others. Applied to such data, the maximum likelihood estimates under panmixia are unbiased, but the variance is underestimated and the population gene frequencies may be some weighted average of two or more generations.

In practice it is rarely necessary to distinguish between gene frequencies of successive generations, and the variances of gene frequency estimates are of little interest, since significance tests are usually inappropriate: the null hypothesis that two groups have the same gene frequencies must be false unless they have the same parents. Therefore, there is no incentive to abandon maximum likelihood theory in finite populations. In particular, formulations in terms of sampling without replacement from a finite population of known gene frequency do not correspond to any practical situation. Problems of gene frequency estimation can always be referred to the infinitely large number of gametes produced by an appropriate (perhaps finite) prior generation.

3.11 Quantitative Phenotypes

We have seen that qualitative phenotypes of individuals give tests of factor-union systems. In principle quantitative phenotypes permit tests of a more general hypothesis: that the overall distribution is a commingling of two or more simple distributions due to different genotypes.

Unfortunately this test requires that we assume particular distributions, taken in the older literature to be normal and with the same variance. Attempts to fit different variances are unsatisfactory because different parameter sets give virtually the same likelihood.

A more useful generalization is to assume that the commingled distributions may not be normal but can be normalized by a simple transformation such as

$$y = \frac{r}{p}\left[\left(\frac{x}{r} + 1\right)^{p} - 1\right] \tag{3.11.1}$$

where $x = (X - \mu)/\sigma$ is the standardized distribution of X in terms of its overall mean μ and standard deviation σ, and both r and p are the same for each distribution. For an arbitrary value of r large enough so that every x in the sample is greater than $-r$ (and therefore $x/r + 1$ is positive), a value of p may be estimated that eliminates skewness within the commingled distributions. If p and r are estimated simultaneously, both skewness and kurtosis, the two main deviations from normality, are controlled, although kurtosis is not eliminated.

The problem with this approach is that there is little reason to assume distributions of this form. The normal distribution ($p = 1$) arises with an infinite number of additive factors of equal effect. The logarithmic normal distribution ($p \to 0$) corresponds to an infinite number of multiplicative factors of equal effect. If the number of causes is not infinite, or if there are both additive and multiplicative effects, the distribution of y in equation 3.11.1 will not be normal. Then this procedure will tend to give a spuriously better fit for commingling of two or more distributions than for a single distribution, just because the assumed family of simple distributions is wrong. Therefore estimation of frequencies of commingled distributions is often misleading. One check is against Hardy-Weinberg proportions in populations where panmixia holds. Significant deviation provides evidence against commingling and in favor of distributional peculiarities. However, a good fit to Hardy-Weinberg proportions does not support commingling, whose extra parameters may be sufficient to fit the observed distribution successfully but erroneously.

If two or more distributions exist, they may have an environmental or technical cause of no genetic interest. Equation 3.11.1 is useful primarily to diminish skewness and kurtosis when that facilitates analysis, following the philosophy of Sewall Wright:

> "We are concerned here with interpretation and with forms that reduce the differences between families of distributions to differences in as few parameters as possible, rather than with the best possible fitting of isolated distributions."

Much of the quantitative variation in red cell acid phosphatase is due to polymorphs at the *ACP1* locus, yet their effects were not suspected until electrophoresis revealed a factor-union system. Thus not only the specificity but the power of tests for commingling is low, and critical evidence comes only from pedigree analysis.

3.12 Questions

1. *In rabbits the albino system has alleles C (wild type), c^h (Himalayan), and c (albino), with the following phenotypic effects:*

$CC = Cc^h = Cc$ = wild type
$c^h c^h = c^h c$ = Himalayan
cc = albino

Draw the phenogram. Is this a factor-union system? If so, how may it be represented?

yes.

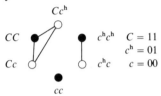

$C = 11$
$c^h = 01$
$c = 00$

2. *There is another factor-union system with the same number of alleles, genotypes, and phenotypes. What is it?*

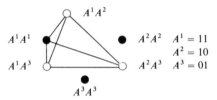

$A^1 = 11$
$A^2 = 10$
$A^3 = 01$

3. *How are the factor pairs in problems 1 and 2 designated?*

   A. Subtypic (10 allele missing, a ⊂ b)
   B. Codominant (00 allele missing)

4. *Let* $q = (1/16, 1/4, 11/16)$ *be a gene frequency vector for alleles* A^1, A^2, A^3. *What is the matrix of genotype frequencies under panmixia? What is the frequency of the genotype* $A^1 A^3$?

$$D = [q_i q_j] = \begin{bmatrix} 1/256 & 1/64 & 11/256 \\ 1/64 & 1/16 & 11/64 \\ 11/256 & 11/64 & 121/256 \end{bmatrix}$$

$2p_{13} = 11/128$

5. *In the above problem, what is the frequency of the mating type* $A^1 A^2 \times A^2 A^3$, *reciprocal matings being distinguished?*

$m_{(12)(23)} = 4(1/64)(11/64) = 11/1024$

6. *What are the frequencies of the matings* $A^1 A^1 \times A^1 A^3$, $A^1 A^2 \times A^1 A^3$, *and* $A^1 A^3 \times A^1 A^3$, *reciprocal matings being pooled?*

$2m_{(11)(13)} = 4(1/256)(11/256)$

$2m_{(12)(13)} = 8(1/64)(11/256)$

$m_{(13)(13)} = 4(11/256)^2$

7. Let a, A be a pair of neutral sex-linked alleles. If a population is started with the mating $aY \times AA$, what are the expected gene frequencies in males and females in the next 4 generations, the tth generation, and at equilibrium?

Generation	Males (q^p)	Females (q^m)
0	1	0
1	0	$1/2 = .5000$
2	$1/2$	$1/4 = .2500$
3	$1/4$	$3/8 = .3750$
4	$3/8$	$5/16 = .3125$
t	$\{1 - (-1/2)^{t-1}\}/3$	$\{1 - (-1/2)^t\}/3$
∞	$1/3$	$1/3 = .3333$

8. A population is started by the mating $aaBB \times AAbb$, where A and B are two systems with recombination frequency .05. What is the expected frequency of ab gametes after 10 generations? How far have the two systems gone toward equilibrium?

$\{1 - (1 - .05)^{10}\}/4 \doteq .1$

$\dfrac{P_{10}(ab)}{P_\infty(ab)} \doteq .4$

9. In how many ways may a family of size 4 with 2 albinos be ordered? Diagram these, using filled circles for albinos and open circles for normal children.

$\dbinom{4}{2} = 6$

● ● ○ ○

● ○ ● ○

● ○ ○ ●

○ ● ● ○

○ ● ○ ●

○ ○ ● ●

10. Letting θ_m, θ_p be the recombination fraction in maternal and paternal

*gametes, respectively, what value should be assigned to θ in equation
3.8.1 for two autosomal systems? For an autosomal and a sex-linked
system? For two sex-linked systems?*

$\theta = (\theta_m + \theta_p)/2$

$\theta = 1/2$

$\theta = 2\theta_m/3$

11. *Suppose that a population includes 300 individuals of a certain genotype, of whom 285 are normal and the rest have a characteristic phenotype. What is the penetrance in this genotype?*

15/300

3.13 Bibliography

Cavalli-Sforza LS, Bodmer WF: The Genetics of Human Populations. Freeman, San Francisco, 1971

Cotterman CW: Factor-union phenotype systems. In: Computer Applications in Genetics.
Edited by Morton NE University of Hawaii Press, Honolulu, 1969, pp 1–18

Crow JF, Kimura M: An Introduction to Population Genetics Theory. Harper and Row,
New York, 1970

Elandt-Johnson RC: Probability Models and Statistical Methods in Genetics. John Wiley,
New York, 1971

Hartl DL: Principles of Population Genetics. Sinauer Associates, Sunderland, 1980

Li CC: Population Genetics. University of Chicago Press, Chicago, 1955

MacLean CJ, Morton NE, Elston RC, Yee S: Skewness in commingled distributions.
Biometrics 32: 695–699, 1976

*Shows TB, Alper CA, Bootsma D, Dorf M, Douglas R, Huisman T, Kit S, Klinger HP,
Kozack C, Lalley PA, Lindsley D, McAlpine PJ, McDougall JK, Meera Khan P,
Meisler M, Morton NE, Opitz JM, Partridge CW, Payne R, Roderick TH, Rubenstein
P, Ruddle FH, Shaw M, Spranger JW, Weiss K*: International system for human gene
nomenclature. Cytogenet Cell Genet 25: 96–116, 1979

Sutton HE: An Introduction to Human Genetics. Second edition. Holt, Rinehart and
Winston, New York, 1975

Wright S: Evolution and the Genetics of Populations II. The Theory of Gene Frequencies.
University of Chicago Press, Chicago, 1969

4. Pairs of Relatives

Whereas individuals give little information about causal hypotheses, except for factor-union systems, pairs of relatives are much more informative. The popularity of pairs of relatives is due to the ease with which data are reduced to pairs, the great economy of this reduction of different pedigree structures to a few correlations or recurrence risks, and the freedom of these estimates under certain assumptions from ascertainment bias.

4.1 Conditional Probabilities

Consider two relatives, X and Y. In the simplest case they are sampled entirely at random and the conditional probability of Y given X is

$$P(Y \mid X) = P(X,Y)/P(X) \tag{4.1.1}$$

For a quantitative trait with bivariate normal distribution (which can for all practical purposes be assured by eq. 3.11.1), this reduces to the univariate normal distribution with mean ρx and variance $1-\rho^2$,

$$P(y \mid x) = N(\rho x, 1-\rho^2) \tag{4.1.2}$$

where ρ is the correlation between X and Y, and $x = (X - \mu_x)/\sigma_x$, $y = (Y - \mu_Y)/\sigma_Y$ are standardized variables in terms of mean μ and variance σ^2. On the assumption of multivariate normality the probability of any pedigree structure consisting of s members can be expressed in terms of such correlations.

A characteristic problem of genetic epidemiology is incomplete selection, which may involve both ascertainment of families and later exclusion of families with no affected members. Analysis is based on the concept of a *proband*, defined as an affected person who, at any time, was detected independently of his relatives, and who would therefore be sufficient to assure selection of those relatives in the absence of other probands. Probands may be ascertained through hospital records, death certificates, inquiries to physicians, examination of a population sample, or other direct means for determining phenotypes. The first proband to be ascertained is called the *propositus* (index case), but serious bias results if other probands are neglected. Affected individuals who are not probands but are detected through family study of probands are called *secondary cases*.

The concept of a proband leads directly to the ascertainment probability. Suppose that in a given finite population there are R affected individuals, of whom A are detected as probands. Then the *ascertainment probability* π is defined as

$$\pi = A/R, \tag{4.1.3}$$

the conditional probability that an affected member of the population be detected as a proband. Under incomplete selection an estimate of π is essential for a valid and efficient analysis of recurrence risks and determination of the number of affected individuals in the population,

$$R = A/\pi \tag{4.1.4}$$

In *truncate selection* $\pi = 1$, and pedigrees with many affected members are no more likely to be ascertained than pedigrees with a single affected member. Truncate is derived from the fact that, if the recurrence risk is constant in a given type of family, then the distribution of the number of affected members is a truncated binomial. Truncate selection is attained by exclusion from a random sample of pedigrees without affected members.

In *single selection* $\pi \to 0$, and the probability that a pedigree be ascertained is nearly proportional to the number of affected members. This is the simplest kind of incomplete selection to analyse in pairs of relatives or pedigrees, but should be avoided whenever practicable for two reasons:
1. When $\pi \to 0$ it is impossible to estimate the number of affected individuals in the population by equation 4.1.4.
2. Single selection gives a poor representation of families with isolated cases which may depend on interesting genetic or nongenetic mechanisms different from familial cases.

The general case is *multiple selection* with $0 < \pi \le 1$. If affection is sometimes so mild as not to seek medical attention or so severe as to suffer premature mortality, then even an intensive investigation over a long period of time may not achieve truncate selection.

Suppose X and Y are two relatives of known affection status, which may be selected for X but not Y. Let m_k be the risk for affection if X belongs to some category k, and f_k be the prior probability of that category, perhaps conditional on the age, sex, or other characteristic of X but neglecting his relatives. Then the *morbid risk* m_x is defined as the probability that X be affected,

$$m_x = \sum_k f_k m_k \tag{4.1.5}$$

Similarly let p_k be the probability that Y be affected, given that X belongs to the kth category. The *recurrence risk* $p_{y \cdot x}$ defined on the pair of relatives (X, Y) is the conditional probability that Y be affected, given that X is affected:

$$p_{y \cdot x} = \frac{\sum_k f_k m_k p_k}{\sum_k f_k m_k} \tag{4.1.6}$$

In the case of a fully penetrant recessive gene, the f_k are the probabilities of the four mating types ($AA \times -$, $Aa \times Aa$, $Aa \times aa$, $aa \times aa$) and m_k are corresponding segregation frequencies (0, $\frac{1}{4}$, $\frac{1}{2}$, 1). More generally, different values of k may denote a new mutant, an environmental factor, a genotype defined at many loci, and so on.

Shortly after Mendelism was rediscovered in the beginning of this century, Weinburg introduced an estimate of recurrence risk in sibships under incomplete selection,

$$p_{y \cdot x} = \frac{\sum a(r - 1)}{\sum a(s - 1)} \tag{4.1.7}$$

where a given sibship has a probands, r affected, and s members $(0 < a \le r \le s)$. It can be shown that this estimate is unbiased but not fully efficient. Fisher gave an efficient estimator

$$p_{y \cdot x} = \frac{\sum Ca(r - 1)}{\sum Ca(s - 1)} \tag{4.1.8}$$

where

$$C = \frac{1}{1 + \pi + p_{y \cdot x} \pi(s - 3)}$$

The effective number of observations is $\Sigma Ca(s-1)$. Sample sizes tend to be exaggerated if the usual formula 4.1.7 is used instead of equation 4.1.8.

Leaving problems of estimating correlations for quantitative data and recurrence risks for affection status, the reader should be aware that good estimates require fastidious methods not used in the older literature.

4.2 Alternative Modes of Inheritance

The number of alternative modes of inheritance which could be postulated is unlimited, but in practice we can only test a few of them. An acceptable model has the three properties of support, economy, and resolution. *Support* means that it gives a good fit to available data as judged by an appropriate criterion like χ^2. *Economy* means that it involves no more parameters than another supported hypothesis, and that the number of residual degrees of freedom is high. *Resolution* means that the available data exclude other plausible models with the same number of parameters. A model is plausible if it has been found to give a good fit with other phenotypes or to the same phenotype in different populations (see fig. 4.2.1).

With attribute data when the phenotype is dichotomized into affected and normal, much of the evidence comes from the population frequency and marital, sib, and parent-offspring pairs. Therefore a maximum of four parameters may be entertained. With quantitative data estimation of an environmental index doubles the number of parameters that may profitably be considered, apart from ancillary parameters of no genetic interest.

Because of these limitations four models have proved useful: (a) the generalized single locus with three parameters, (b) additive cultural inheritance with three to six parameters, (c) additive polygenic inheritance with two parameters, and (d) the mixed model including two or more of the above.

Complications ignored by these models include metataxy, parataxy, multiple alleles, and nonadditivity of multiple loci. These complexities would make different models nonresolvable. The utility of simpler models is to: (a) fit data economically, (b) resolve genetic and cultural inheritance, (c) provide predictions for genetic counseling under acceptable hypotheses, (d) resolve major loci from continuous variation, and (e) resolve etiological heterogeneity. Experience has suggested that pedigree analysis is essential for the last two objectives, but that pairs of relatives are useful to meet objectives (a) to (c).

Discontinuous Variation

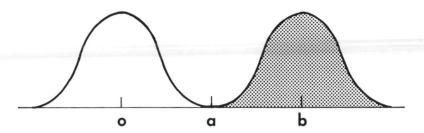

Continuous Variation

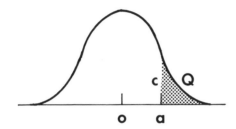

Mixed Model

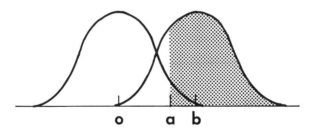

Fig. 4.2.1: Models of discontinuous and continuous variation.

4.3 The Generalized Single Locus

This assumes two alleles at an autosomal locus, say A, A' (see table 4.3.1). On a scale (which may be an observed quantitative trait or conceptual "liability" to affection) the means of the three genotypes may be expressed in terms of three parameters:

Table 4.3.1 : The generalized single locus model

Genotype	AA	AA'	$A'A'$
Frequency	$(1-q)^2$	$2q(1-q)$	q^2
Mean	z	$z + dt$	$z + t$

z, the mean of AA
t, the displacement between homozygotes
d, the degree of dominance $(0 \leq d \leq 1)$

If the effects of A and A' are additive $(d = \frac{1}{2})$, we say that there is no dominance. Assuming panmixia and neglecting mutation and selection, the population genotype frequencies are given by q, the gene frequency of A'. Transmission is mendelian: i.e., the transmission frequency $\tau(A')$ of A' gametes from parents of genotypes AA, AA', $A'A'$ is 0, 1/2, 1. If there are discrete risk classes defined by age, sex, or other observed variables, these effects are taken to be additive with the major locus. Then the ith class has homozygous mean z_i, where i is called the *liability indicator*. With quantitative data the assumption of additivity allows a continuous *liability index* to be incorporated into the definition of the phenotype by covariance analysis.

Besides the major locus, there is an additive effect of random environment. Thus multiple loci and cultural inheritance are neglected. In symbols

$$x = g + e$$

where x is liability or a covariance-adjusted quantitative trait, g is the major locus effect, and e is random environment. For the ith liability indicator g takes only the values z_i, $z_i + dt$, $z_i + t$, and e has a normal distribution with mean zero and variance E. The mean of g in liability class i is

$$\mu_i = z_i + qt\{2 + q(1 - d)\}$$

with variance $G_i = q^2(z_i + t)^2 + 2q(1 - q)(z_i + dt)^2 + (1 - q)^2 z_i^2 - \mu_i^2$.
On these assumptions, G_i is constant (say G) for all i. The total variance V within liability class is $V = G + E$. Since V and the μ_i are determined in the data, and z_i may be substituted for in terms of μ_i, d, t, and q, the model is specified by the three genetic parameters d, t, and q. For a liability scale the variance V is arbitrarily taken to be unity, and one z_i is set to zero.

A serious limitation of the single locus model is that it provides no specific test for other genetic or cultural inheritance. It is a special case of the mixed model, which may be tested in pedigrees but is often indeterminate with pairs of relatives. A good fit of the single locus model to pairs of relatives obviously should not be taken as proof, and should lead to collection and analysis of critical pedigree data. Meanwhile the estimates may be used cautiously in genetic counseling. If on the other hand pairs of relatives do not fit a single locus, the incentive to examine pedigree data is no less, but provisional use of estimates in genetic counseling is more questionable.

Unfortunately, for most diseases the quality and quantity of data on pairs of relatives are modest. Quality is degraded by inaccurate definition of probands and inefficiency of risk estimation. Quantity includes both sample sizes and diversity of relationships. Pending analysis of better data, inference can only be tentative. The genetic epidemiologist must try to extract all the information in the data, without exaggerating the evidence.

4.4 Genetic and Cultural Inheritance

For many traits there is no suggestion of a major locus, and the randomness of environment among families is doubtful. Without abandoning the search for major loci by appropriate pedigree analysis of more precisely defined phenotypes, it is tempting to consider the linear model

$$x = g + b + e \qquad\qquad (4.4.1)$$

where x is liability or a covariance-adjusted quantitative trait, g is the contribution of additive genetic factors (polygenes) with mean zero and variance G, b is the contribution of family environment with mean zero and variance B, and e is the contribution of random environment with mean zero and variance E. All three components are taken to be normally distributed and (for most phenotypes) uncorrelated. The total variance V is G + B + E. In this model G represents genetic inheritance and B is cultural inheritance. It is a special case of the mixed model, with a major locus eliminated to give determinacy in pairs of relatives.

Linear models lend themselves to path diagrams in which arrows go from cause to effect and the path coefficients represent standardized partial regression coefficients. Unmeasured variables are identified by ellipses and measured variables by rectangles.

In figure 4.4.1 genetic inheritance involves the parameter h and the

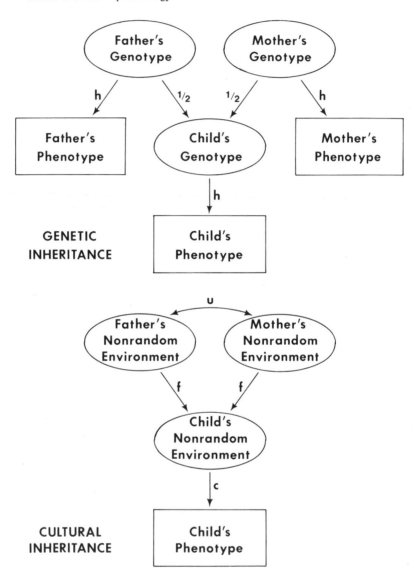

Fig. 4.4.1: A simple model for genetic and cultural inheritance.

coefficient 1/2. *Genetic heritability* h^2 is the proportion of the variance due to additive genetic factors. In terms of equation 4.4.1, $h^2 = G/V$. For this simple model heritability is assumed the same in parents and children, as if they were observed at the same age (table 4.4.1). Parental

Table 4.4.1: Determinants of systolic blood pressure in different populations (Although constancy of parameters is not expected, there is reasonable agreement among studies)

Population	Index[a]	Genetic heritability (h^2)	Cultural heritability (c^2)
Japanese-American	+	0.247	0.156
Tokelau adults	+	0.316	0.050
Tokelau children	+	0.224	0.009
Brazil	+	0.306	0.042
Literature	0	0.254	0.072
Family sets	0	0.194	—

[a] Use of environmental indices is indicated by +

genotypes are assumed to be uncorrelated. The coefficient 1/2 signifies that half of the child's genes came from each parent, and therefore that the correlation between genotypes of child and one parent is 1/2. Only $(1/2)^2 + (1/2)^2 = 1/2$ of the child's genotype is explained by the two parental correlations, the remainder being due to accidents of segregation in parental gametes, each of which contains only 1/2 of the diploid genes.

Cultural inheritance involves the parameters u, f, and c. *Nonrandom environment* is that part of the environment which may be correlated between family members, the marital correlation being u. The child's nonrandom environment is determined partly through parental nonrandom environments, the parameter f being the cultural analog of 1/2 for genetic inheritance. A proportion $2f^2(1 + u)$ of the variance of child's nonrandom environment is determined by the parents. More generally, there may be different determination by father's and mother's nonrandom environment through paths f_F and f_m respectively. *Cultural heritability*, the proportion of the variance of child's phenotype determined by the nonrandom environment, is $c^2 = B/V$.

The above results depend on two rules of path analysis:
1. Causes are denoted by single-headed arrows directed to effects. Two-headed arrows are used for unanalyzed correlations. Let p_{ij} denote a causal path to i from j, and r_{jk} be a correlation between j and k. Then the simple correlation r_{ik} between i and k is

$$r_{ik} = \sum_j p_{ij} r_{jk}$$

A corollary is that causal paths are never traced through arrowheads incident to the same cause.

2. Let p_{iu} be a residual path due to all factors acting on i not included in the path diagram, and not correlated with them. Then there is complete determination and

$$\sum_j p_{ij} r_{ij} + p_{iu}^2 = 1 \qquad (4.4.3)$$

Random environment is an example of a residual path. Usually residual paths are not included in path diagrams, since by equation 4.4.3 they are determined by

$$p_{iu} = \sqrt{1 - \Sigma_j\, p_{ij} r_{ij}} \qquad (4.4.4)$$

In genetic applications the logical limits are

$$0 \le p_{ij} \le 1$$
$$-1 < r_{ij} \le 1 \qquad (4.4.5)$$

for all i, j.

Path analysis was created by Sewall Wright, one of the three great population geneticists, as an aid in analysis of causation, especially with nonexperimental data such as characterize genetic epidemiology. Haldane considered that path analysis "may replace our old notions of causation." It is most usefully regarded as a way of deriving consequences of linear causal assumptions in terms of an economical set of causal parameters, and then testing these assumptions on correlation structures such as pairs of relatives. Path analysis is a special type of multiple correlation admitting errors in the independent variables, and so the likelihood ratios of correlation theory provide tests of alternative causal models.

4.5 Assortative Mating

For most phenotypes of interest to genetic epidemiology, the possibility of assortative mating is remote. Potential mates do not know one another's genotypes, liability to disease, or chemical compositions, and so spouse correlations for such traits are likely to have postmarital causes such as common diet, habits, and exposure to disease. For traits involving significant errors of measurement, such as blood pressure or serum cholesterol, the usual study designs may create spurious correlations due entirely to the fact that family members are tested by the same observer at the same time. Genetic epidemiologists should make more effort to avoid such observer effects by randomizing times and observers, and should be cautious when interpreting older data where these precautions were neglected. For some behavioral and anthropometric characters substantial marital correlation has been found (roughly .5 for IQ and height), which is partly due to premarital assortment or *homogamy*. In such cases one mate obviously does not cause the other, nor does one sex choose a completely nonselective partner. We need a special symbol and calculus for mutual selection. Cloninger introduced the co-path, designated by a headless arrow, for a partial correlation specific to a pair of causes, independent of other factors. He showed that a co-path should be traced in either direction without regard to whether incident variables are causes or effects, and is therefore logically different from the single and double-headed arrows of classical path analysis.

Figure 4.5.1 represents assortative mating in terms of a co-path *p* between marital phenotypes (representing *phenotypic homogamy*) and a correlation *u* between marital nonrandom environments (representing *social homogamy*). Phenotypic homogamy is always premarital, whereas social homogamy may be premarital, postmarital, or both. The cause of premarital social homogamy is division by social class, religion, geography, or other factors into somewhat endogamous groups with different

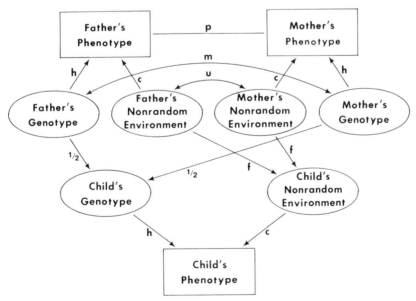

Fig. 4.5.1: Assortative mating with phenotypic homogamy (*p*) and social homogamy (*u*, *m*).

patterns of life. Postmarital social homogamy may be due to cohabitation or observer bias or both.

In principle social homogamy could induce a correlation m between marital genotypes, if premarital grouping were racial, if social classes were genetically differentiated, or if the potential mates belong to clubs or institutions for particular diseases. So far a significant value for m has not been detected, perhaps because conspicuous racial and environmental diversity is usually avoided in studies of pairs of relatives.

4.6 Intergenerational and Sex Differences

The causal scheme in figure 4.6.1 was simplified by omission of intergenerational and sex differences. If (as is nearly always the case) parents and children are observed at different ages, the measured traits are not the same. To represent this we must distinguish between the genetic path to children (h) and adults (hz) and the corresponding cultural paths (c and cy respectively), with logical constraints $y, z > 0$ (table 4.6.1).

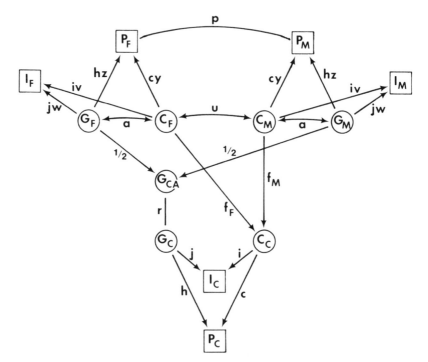

Fig. 4.6.1: Family resemblance with assortative mating, indices, maternal effects, and intergenerational differences.

Table 4.6.1 :   Parameters of the mixed homogamy model

Symbol	Definition
h	Effect of genotype on child's phenotype (square-root of heritability)
hz	Effect of genotype on adult's phenotype
c	Effect of child's indexed environment on the child's phenotype
cy	Effect of adult's indexed environment on the adult's phenotype
p	Primary correlation between parental phenotypes, not due to secondary resemblance through social homogamy
m	Correlation between parental genotypes through social homogamy
u	Correlation between parental indexed environments through social homogamy
f_F	Effect of father's indexed environment on child's indexed environment
f_M	Effect of mother's indexed environment on child's indexed environment
b	Effect of nontransmitted common sibship environment on child's indexed environment
bx	Effect on nontransmitted common sibship environment on adult's indexed environment
i	Effect of child's indexed environment on the child's index
iv	Effect of parent's indexed environment on parent's index
j	Effect of child's genotype on child's index
jw	Effect of parent's genotype on parent's index
r	Co-path between genotypes of child and adult

A more subtle intergenerational difference is that the path from parental to child's genotype will fall short of 1/2 if different genes act in children and adults. We introduce a co-path r between the genotypes which the same individual expresses as child and adult. This complexity need not be entertained for cultural inheritance, being confounded with the parameter f from parent's to child's nonrandom environment.

The environment and constitution of the mother may act on her fetus or infant as a *maternal effect* with $f_M > f_F$, where subscripts M and F denote mother and father, respectively. There does not seem to be any loss of generality in this formulation. For example, an unobserved aspect of the maternal constitution could have $y \to 0$, but possibly $f_M \to 1$.

Unsuccessful attempts have been made to introduce paths from phenotypes of parents to children. Most data structures cannot resolve such paths from genetic and cultural ones. Even for IQ the phenotypic path is nonsignificant. There is a logical inconsistency in considering current parental phenotype, imperfectly measured, as a cause of child's past development.

Sibs often share environment which was not transmitted from parents. This is represented by b for sibs as children and by bx for sibs as adults.

4.7 Quantitative Traits

A continuous randomly-sampled variable must usually be prepared for analysis by *covariance-adjustment* for age, sex, and other factors considered extraneous to family resemblance, perhaps after preliminary transformation (logarithmic, square root, etc.) to reduce skewness. Outlying observations likely to be measurement errors should be rejected. The underlying model for regression is

$$X = A + \Sigma b_i Z_i + \varepsilon \tag{4.7.1}$$

where X is the dependent variable, Z_i is an independent variable which might be age, sex, sex $\times$ age^3, and so forth, and ε is normally distributed error with mean 0 and variance σ^2. Then

$$x = (X - A - \Sigma b_i Z_i)/\sigma \tag{4.7.2}$$

is a covariance-adjusted value of X. If the preliminary transformation did not stabilize variance, the appropriate adjustment is

$$x = (X - A - \Sigma b_i Z_i)/\sigma_x \tag{4.7.3}$$

where σ_x is the standard deviation in the class defined by $A + \Sigma b_i Z_i$ to which x belongs. A final power transform as in equation 3.11.1 will eliminate skewness and reduce kurtosis. The purpose of this manipulation of the data is to generate a phenotype from which predictable effects of extraneous factors have been removed and with a nearly normal distribution in random samples.

Sometimes a quantitative trait is selected through probands with extreme values. This ascertainment bias should be removed. With pairs of relatives it is best handled by regression on probands, using the principle that

$$r_{ij} = b_{ij}(\sigma_j/\sigma_i) \tag{4.7.4}$$

where the regression coefficient b is not biased by selection of probands, and σ_i, σ_j are the standard deviations in the population. Multiple probands create estimation problems that have not been completely resolved.

After covariance adjustment and power transformation, a quantitative trait P can usually be predicted by other variables z_i representing such aspects of life style as social class, physical exercise, diet, smoking, or alcohol consumption. By multiple regression we may obtain

$$P = a + \Sigma b_i z_i + \varepsilon' \tag{4.7.5}$$

and take $I \equiv \Sigma b_i z_i$ as an *index* of P. Figure 4.6.1 shows a path diagram which includes indices as well as other complexities. Subscripts F, M, C denote father, mother, child, respectively. The subscript CA stands for child as an adult. The phenotype P and index I form a pair of observations partly determined by genotype G and nonrandom environment C, together with residual paths for random environment, which are not shown. Intergenerational differences are included for paths to phenotypes (y, z) and indices (v, w). If the gene-environment correlation a is constant for parents and children, it is not an independent parameter but a function of the other paths.

Although this model seems complicated, it is fully determined in nuclear families consisting of parents and children. Unusual relationships like adopted children, identical twins, half-sibs, children of identical twins, and more remote relatives provide additional information. The genetic epidemiologist may choose to concentrate on indices in nuclear families, since they are the most significant relationship. Alternatively, he may neglect indices and collect data on unusual or remote relationships, where special factors like social bonding of twins or selective placement of adopted children may disturb extrapolation to family resemblance in usual situations. The wisest course seems to take nuclear families with indices as the primary object of study, with unusual and remote relationships providing an additional test of the model.

Ideally the data would be collected by the same investigator in a single population. In practice there is much variation among studies and populations. When different studies are pooled, the variance among replicates should be used as error in an F test, instead of relying on an inappropriate χ^2 test of goodness of fit.

Estimates of variance components from this model frequently show intergenerational differences and maternal effects. Cultural inheritance is often present, and sometimes is a more important determinant of family resemblance than genetic inheritance. An appropriate measure of the genetic component in family resemblance is

$$
R_G = \begin{cases} \dfrac{h^2/2}{c^2 + h^2/2} & \text{for children} \\[2ex] \dfrac{h^2 z^2/2}{c^2 y^2 + h^2 z^2/2} & \text{for adults} \end{cases} \tag{4.7.6}
$$

since all of cultural heritability but only half of genetic heritability contributes to family resemblance.

4.8 Affection Status

A quantitative trait X may be reduced to a dichotomy by defining a threshold Z such that $X < Z$ is called "normal" and $X > Z$ is "affected". If X is normally distributed with mean μ and variance σ^2, the affection probability is

$$P(\text{aff} \,|\, X) = \frac{1}{\sqrt{2\pi}} \int_z^\infty e^{-x^2/2} \, dx \equiv Q(x) \tag{4.8.1}$$

where $x = (X - \mu)/\sigma$ and $z = (Z - \mu)/\sigma$. With a computer $Q(x)$ may be calculated to any degree of accuracy.

Much information is lost when a quantitative trait is replaced by affection status, since we do not know whether an individual is close to or far from the threshold. This loss of information outweighs the convenience of not having to worry about deviation of the quantitative trait from normality. Moreover, we can no longer remove extraneous variation by covariance analysis or create an index linear on X.

Instead, a discriminant may be formed by regression of affection status (as a 0, 1 variable) on age, sex, and other relevant factors. Since a quantitative discriminant is nonlinear on X, it is convenient to polychotomize the discriminant into a small number of classes, $i = 1, \ldots n$, with increasing affection risk a_i. Then i is called a *liability indicator*. This allows us to write an explicit expression for

$$P_{jk\ell} \equiv P(\text{Ego}_j = \text{affected} \,|\, \text{relative}_{k\ell} = \text{affected}), \tag{4.8.2}$$

where the subscripts j, k denote liability indicators and ℓ is a type of relationship. Such a probability is a complex function of ℓ, a_j, a_k, and the correlation of the quantitative trait (tables 4.8.1 and 4.8.2). Treating pairs of relatives as independent, their likelihood is

$$L = \prod_{j, k, \ell} P_{jk\ell}^{r_{jk\ell}} (1 - P_{jk\ell})^{c_{jk\ell}} \tag{4.8.3}$$

where $r_{jk\ell}$ is the number of affected and $c_{jk\ell}$ the number of normals in the set defined by j, k, ℓ.

This theory is numerically difficult but simple conceptually. It is appropriate whether a quantitative trait was observed and dichotomized or merely an abstract liability scale on which the various components of a model act linearly. Indicator-specific penetrance for each major locus genotype is predicted from such a model, and need not be estimated independently which is rarely feasible.

Table 4.8.1: Observed and expected numbers of definite schizophrenics among nine types of relatives (population incidence = 0.85%)

Type of relatives	Expected correlation	r^a	Sample size	Definite schizophrenics		
				Obs	Exp	χ^2
Spouses	$c^2 u$	.134	194	4	4.70	0.10
Children	$\dfrac{h^2}{2} + c^2 f(1 + u)$	.482	1578	178	159.03	2.51
Sibs	$\dfrac{h^2}{2} + 2c^2 f^2(1 + u)$	.410	8817	736	734.83	0.00
MZ twins	$h^2 + 2c^2 f^2(1 + u) + t^2$	.854	261	119	118.61	0.00
DZ twins	$\dfrac{h^2}{2} + 2c^2 f^2(1 + u) + t^2$	.530	329	45	39.90	0.74
Half sibs	$\dfrac{h^2}{4} + 2c^2 f^2(1 + u)$	.222	499	17	17.97	0.05
Nieces and nephews	$\dfrac{h^2}{4} + 2c^2 f^3(1 + u)^2$	.176	3966	105	122.90	2.69
First cousins	$\dfrac{h^2}{8} + 2c^2 f^4(1 + u)^3$	.089	1600	25	27.27	0.19
Grandchildren	$\dfrac{h^2}{4} + c^2 f^2(1 + u)^2$	.188	739	21	25.81	0.92

Goodness-of-fit test (Pearsonian): $\chi_4^2 = 7.20\,(P > .12)$
^{a} These are tetrachoric correlations estimated from the observed data

More information can be extracted from affection status if three or more phenotypes are recognized. In a trichotomy the third class may either be intermediate between normal and affection, or include affection. The first use corresponds to M < X < Z in figure 4.8.1, and the second case is X > M, where Z is the threshold for affection and M the threshold for intermediacy. It would be natural to take the exclusive definition if X is observed and trichotomized, or more generally if intermedi-

Table 4.8.2: Tests of hypotheses and estimates of parameters for schizophrenia

Hypothesis	χ^2	df	h^2	c^2	t^2	f	u
General	7.20	4	.707	.203	.090	.277	.790
No genetic heritability	128.94	5	0	.718	.282	.481	.207
No cultural heritability	28.31	7	.830	0	.039	0	0
No environment unique to twins	19.83	5	.814	.186	0	.151	.856
No assortative mating	14.74	5	.711	.207	.082	.376	0

ates are of different phenotype from affected. The inclusive definition is appropriate if some intermediates are affected.

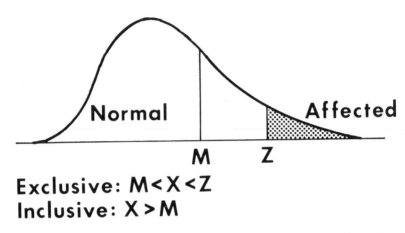

Normal   **Affected**

M   Z

Exclusive: M < X < Z
Inclusive: X > M

Fig. 4.8.1: Two definitions of a trichotomy. The intermediate class may either exclude or include affection.

4.9 Questions

1. *If 25 probands are observed in a population of 1 million and the ascertainment probability is .1, what is the trait frequency?*

$$P = \frac{25}{(.1)10^6} = .00025$$

2. *For a certain disease the population frequency and the risk in first degree relatives have been estimated, but there is no secure information about mode of inheritance. What model would a prudent genetic counselor use to estimate risk for a nephew of an affected individual?*

The polygenic model with no difference between generations. The predicted risk would have to be accompanied by a clear statement of its slender basis.

3. *Given data on population frequency, risk in children, and risk in sibs, could the generalized single locus and polygenic inheritance be resolved? Why?*

No. The difference between risks in children and sibs would be accounted for by dominance in the first case, and by intergenerational difference (z) in the second.

4. *If* $p = m = 0$ *in figure 4.5.1, what fraction of the child's nonrandom environment is determined by parental nonrandom environment?*

   $2f^2(1 + u)$

5. *It is sometimes claimed that height, being measured on a linear scale, is thereby more suitable for analysis of family resemblance than the intelligence quotient (IQ), which is based on a contrived scale of normalized response. Is this true?*

   No. For genetic analysis both must be standardized to eliminate skewness and make variance uniform with age. After such transformation, one trait is no more "natural" or "valid" than another. The precision (repeatability) of height measurement is much greater than for IQ, but this is not dependent on linearity of scale.

6. *Covariance adjustment for age is rejected by some genetic epidemiologists because displacement of major locus genotypes may change with age. Suggest an alternative approach.*

   A different power transform may be used for parents and children.

7. *Suppose for a particular trait that the power transform which eliminates skewness after age and sex adjustment has been found consistent in different populations, but age and sex effects are variable. A pedigree sample is collected based on probands with elevated values of the trait, but no control sample is reported. Suggest a reasonable method to analyze such data.*

   Form the regression

   $$X = A + \Sigma b_i Z_i + BRX_p + \varepsilon$$

where the Z_i are terms in age and sex, X_p is the X value of a proband, and R is the relationship to the proband: ie, $R = 1/2^n$ for n^{th} degree relatives.

   Take $x = (Y - A - \Sigma b_i Z_i)/\sigma_x$ as the variate to be power-transformed, where σ_x is the standard deviation for a class of $A + \Sigma b_i Z_i$.

8. *From Student's distribution, what would be a reasonable lower limit to the sample size on which an estimate of σ_x is based?*

The choice is somewhat arbitrary, but 100 would be a reasonable lower limit. An alternative approach would be to estimate σ_x as a continuous function of $A + \Sigma b_i X_i$, but this often gives unreliable results for extreme values.

9. *Five independent properties are attributed to nonrandom environment in figure 4.6.1. What are they, and how are they symbolized?*

Vertical transmissibility (f), contagion to sibs (b), efficacy (c), estimability (i), and linearity.

10. *What are the properties of random environment?*

No correlation with any other cause, and therefore nonestimability.

11. *The conventional discriminant formed by regressing affection status (0, 1) on other factors is not linear on liability. Is this undesirable?*

No. The discriminant is used to define a discontinuous liability indicator with estimated affection rates. Linearity on liability is not assumed.

12. *How does gene-environment covariance enter analysis of family resemblance?*

Through the homogamy parameters p, u, and m.

13. *How does gene-environment interaction affect analysis of family resemblance?*

Interaction with random environment contributes only to the residual variance, which includes errors of measurement and is not explained by analysis of family resemblance. Interaction with nonrandom environment contributes to c and h proportionately. As far as is known, equation 4.7.6 is not altered. A good fit to an additive model increases confidence that interactions are negligible.

14. *How does gene interaction (epistasis) affect analysis of family resemblance?*

A fraction inflates heritability. Most epistasis is confounded with random environment and environment common to sibs.

4.10 Bibliography

Cloninger CR: Interpretation of intrinsic and extrinsic structural equations by path analysis. Theory and applications to assortative mating. Genet Res 36: 133–145, 1980

Haldane JBS: A defense of beanbag genetics. Perspect Biol Med 7: 343–359, 1964

Li CC: Path Analysis, The Boxwood Press, Pacific Grove, 1975

Morton NE, Rao DC: Quantitative inheritance in man. Yearbk Phys Anthrop 21: 12–41, 1978

Reich T, James JW, Morris CA: The use of multiple thresholds in determining the mode of transmission of semi-continuous traits. Ann Hum Genet 36: 163–184, 1972

Rao DC, Morton NE, Cloninger CR: Path analysis under generalized assortative mating. I. Theory. Genet Res 33: 175–188, 1979

Rao DC, Morton NE, Lew R: Path analysis of attribute data on pairs of relatives. Application to schizophrenia. Hum Hered 1980, in press

Wright S: Evolution and the Genetics of Populations. I. Genetic and Biometric Foundations. University of Chicago Press, Chicago, 1968

5. Pedigrees

By exploiting indices and unusual relationships, pairs of relatives can resolve genetic and cultural inheritance. Pedigrees offer no special advantage for this purpose. Their unique value is to resolve major loci from continuous variation. In this context major means recognizable, which usually requires that the effect be *megaphenic*: ie, greater than one standard deviation in the population.

5.1 Inspection of Pedigrees

Simple modes of inheritance with complete penetrance can be recognized by inspection of pedigrees, using the following criteria:

1. *Autosomal dominant.* (a) Transmission continues from generation to generation, without skipping; (b) except for mutants, every affected child has an affected parent; (c) in marriages of an affected heterozygote to a normal homozygote, the segregation frequency is $\frac{1}{2}$; and (d) the two sexes are affected in equal numbers.

2. *Autosomal recessive.* (a) If the trait is rare, parents and relatives, except siblings, are usually normal; (b) if the recessive genes are alleles, all children of two affected parents are affected; (c) in marriages of two normal heterozygotes, the segregation frequency is $\frac{1}{4}$; (d) the two sexes are affected in equal numbers; and (e) if the trait is rare, parental consanguinity is elevated.

3. *Sex-linked dominant.* (a) Heterozygous mothers transmit to both sexes in equal frequency with a segregation frequency of $\frac{1}{2}$; (b) hemizygous, affected males transmit the trait only to their daughters, the segregation frequency being 1 in daughters and 0 in sons; (c) except for mutants, every affected child has an affected parent; and (d) If the trait is rare, its frequency in females is approximately twice as great as in males.

4. *Sex-linked recessive.* (a) If the trait is rare, parents and relatives, except maternal uncles and other male relatives in the female line, are usually normal; (b) hemizygous, affected males do not transmit

the trait to children of either sex, but all their daughters are heterozygous carriers; (c) heterozygous, carrier women are normal but transmit the trait to their sons with a segregation frequency of $\frac{1}{2}$, and $\frac{1}{2}$ of the daughters are normal carriers; (d) excluding XO daughters of carrier mothers, affected females come only from matings of carrier females and affected males, and their frequency in the population is approximately the square of the frequency of affected males; and (e) except for mutants, every affected male comes from a carrier female.

Other modes of inheritance have been claimed in man, including Y-linkage, partial sex linkage, attached-X chromosomes, and cytoplasmic inheritance, but the evidence for them is inadequate. Y-linkage leads exclusively to father-son transmission, but the only proven Y-linked traits in man have been revealed by cytogenetics.

Some traits are limited in expression to one sex, although members of the opposite sex may transmit a genetic factor for them. Such characteristics are called sex-limited and include premature pattern baldness, imperforate vagina, and other traits associated with reproductive physiology or sex differentiation. If the gene is recessive, all the criteria of this mode of inheritance hold, except that only one sex is at risk. If the gene is an autosomal dominant and limited to males, there is a possibility of confusion with recessive sex-linkage. In both cases, normal carrier females have a segregation frequency of $\frac{1}{2}$ among sons and 0 among daughters. However, the following differences are apparent:

1. An affected, heterozygous male has a segregation frequency of $\frac{1}{2}$ in his sons under autosomal sex-limitation, but 0 under sex linkage.
2. A sex-linked gene may show reduced recombination with other sex-linked genes, while an autosomal gene can show reduced recombination only with other autosomal genes.
3. Through nondisjunction, a carrier for a sex-linked recessive may produce an affected XO daughter, recognizable both clinically and cytologically.
4. If the gene is rare, the frequency of sporadic mutants among all affected cases is $m/(m + 1)$ for an autosomal sex-limited gene and $mu/(2u + v)$ for a sex-linked recessive, where m is the selection coefficient against affected males and u and v are the mutation rates in egg and sperm respectively. For a nearly lethal gene, m approaches 1 and $m/(m + 1)$ approaches $\frac{1}{2}$. Consequently, a frequency of sporadic cases significantly less than $\frac{1}{2}$ is evidence for sex-linkage.

As penetrance decreases or if there is differential mortality before diagnosis, these simple modes of inheritance become indistinguishable to the eye from polygenic and cultural transmission. We need statistical methods to estimate parameters and test hypotheses about the frequencies with which a trait appears in different types of matings, which are the basic data of formal genetics.

5.2 Segregation Frequencies

The mechanisms of inheritance have as their end result the generation of phenotype frequencies characteristic of a given mating type in a specified environment. We shall refer to these as segregation frequencies, including not only classical mendelian frequencies like 1/2, 1/4, and 9/16, but also modifications by differential mortality and partial manifestation, and even empirical frequencies, the genetic basis of which is unclear. For example, if 1/4 of the children from a particular mating type are expected to be of a certain genotype, but only 80 percent of them develop a characteristic abnormal phenotype, we will say that the segregation frequency is $(1/4)(.80) = .20$. Similarly, if 20 percent of the children from a particular mating type have a certain phenotype, we shall say the segregation frequency is .20, even though the mechanism of inheritance may be unknown. Although segregation frequencies are specified by hypothesis, or empirically by analysis of a series of matings, the actual proportions observed in any particular sample are dependent on gene frequencies, chance, the way the data are collected, and other factors.

For simplicity we shall consider only two segregant phenotypes, which may be called *normal* and *affected*. This loses no generality, since n phenotypes may be examined pairwise in $n - 1$ independent ways. Let there be r affected children in a sibship of size s (the terms *sibship, family,* and *mating* will be used interchangeably to denote the set of children classified as normal or affected). A mating will be called *nonsegregating* if there are no affected children ($r = 0$); *segregating* if there is at least one affected child ($r > 0$); and *doubly segregating* if there are both normal and affected children ($0 < r < s$). The *segregation frequency p* is the expected proportion of affected children in a given mating type under complete selection, if this proportion is considered to be uniform or continuously variable. However, if the expected proportion of affected children is considered to be sharply discontinuous, with a low value in some matings and a higher value in others, p will be used to denote the segregation frequency in the latter *high-risk* group of families. We shall naturally be concerned to determine which model is appropriate in any particular body of data.

If the segregation frequency p is constant, with no families that cannot segregate, the distribution of r affected children in a family of size s is binomial.

$$P(r; s,p) = \binom{s}{r} p^r (1 - p)^{s-r} \qquad 0 < p < 1, 0 < r < s \tag{5.2.1}$$

This and other distributions encountered in segregation analysis are most easily and efficiently studied by maximizing the likelihood.

In many cases the segregation frequency p is constant, but there is a proportion h of families that cannot segregate and are indistinguishable from the potentially segregating families which with probability $(1 - p)^s$ happen to produce no affected children. Then the distribution of r affected children in a family of size s is an augmented binomial,

$$P(r; h,s,p) = \begin{cases} h + (1 - h)(1 - p)^s, & r = 0, \quad 0 < p, h < 1, 0 \le r \le s \\ (1 - h)\binom{s}{r} p^r (1 - p)^{s-r}, & r > 0 \end{cases} \tag{5.2.2}$$

The three principal reasons for families that cannot segregate are homozygosity, phenocopies, and bivalent alleles. The probability of homozygosity for a dominant allele can be calculated from the gene frequencies assuming panmixia. For example, consider two alleles (a,A) with frequencies q and $1 - q$. A mating $A- \times aa$ cannot segregate if the $A-$ parent is AA. This conditional probability is

$$h = \frac{P(AA)}{P(AA) + P(Aa)} = \frac{1 - q}{1 + q}. \tag{5.2.3}$$

Similarly, a mating $A- \times A-$ can segregate only if *both* parents are Aa. The conditional probability that a mating cannot segregate is

$$h = 1 - \left[\frac{P(Aa)}{P(AA) + P(Aa)}\right]^2 = \frac{(1 + 3q)(1 - q)}{(1 + q)^2}. \tag{5.2.4}$$

Corresponding probabilities for multiple alleles are easily worked out. Yasuda showed that values of h are not appreciably affected by moderate departures from panmixia (ie, both q and $1 - q$ much greater than inbreeding).

A second important reason for families unable to segregate is phenocopies. A *phenocopy* is a nongenetic affection which mimics a genetic condition closely enough to be sometimes misclassified, although at the molecular level quite distinct. For example, X-radiation during the first three months after conception can produce nonheritable microcephaly with mental retardation similar to the autosomal recessive form. Retinoblastoma is sometimes due to a rare dominant gene with incomplete penetrance and sometimes to a phenocopy that may be a somatic mutation. Severe mental defect or deaf mutism due to intrauterine infection may not be recognized as such.

Some phenocopies are technical errors. For example, in the mating MN × M, the probability of all MN children may be taken to be $h + (1 - h)(1 - p)^s$, where h is the probability that a parent of phenotype MN be actually of genotype NN but give a false positive reaction with anti-M serum. This distribution has been used to argue that technical errors are not frequent enough to explain the excess of MN children observed in published family studies, which therefore may be due to preferential survival of MN fetuses or even meiotic drive. However, this argument is not decisive, since two large studies did not show MN excess.

The third possibility for inability to segregate is a *bivalent* allele determining two factors that often occur in repulsion. Examples of rare bivalent alleles confirmed by finding $h > 0$, $p = 1/2$ are $Gm^{1,3}$ in populations of Asian origin and $Rh^{C,c}$ in populations of African origin.

5.3 Incomplete Ascertainment

Rare traits are often studied in families selected through affected parents or children. Parental selection presents no problem so long as probabilities are written conditional on parental phenotypes. Selection of families through children, with exclusion of nonsegregating families, is called *incomplete selection*. It poses a serious problem for segregation analysis.

The simplest case is a constant ascertainment probability π that an affected member of a population be detected as a proband (see section 4.1). On this assumption the probability of a probands among r affected is binomial

$$P(a; r,\pi) = \binom{r}{a}\pi^a(1 - \pi)^r. \tag{5.3.1}$$

Therefore the probability that a family with r affected have at least one proband is

$$P(a > 0; r,\pi) = 1 - (1 - \pi)^r. \tag{5.3.2}$$

If the segregation frequency p is constant, the probability that a family with s children be ascertained is

$$P(a > 0; s,p,\pi) = \sum_{r=0}^{s} \binom{s}{r} p^r (1-p)^{s-r} [1 - (1-\pi)^r] \tag{5.3.3}$$

$$= 1 - (1 - p\pi)^s.$$

Note that $p\pi$ is the probability that a child be affected *and* a proband, $(1 - p\pi)^s$ is the probability that none of s children be a proband, and therefore $1 - (1 - p\pi)^s$ is the probability that a family with s children be ascertained through the children. A mathematician would say that $1 - (1 - p\pi)^s$ is the measure of ascertained families of size s. We may now evaluate the conditional probability

$$P(r \mid a > 0; s,p,\pi) = \frac{P(r; s,p) \cdot P(a > 0; r,\pi)}{P(a > 0; s,p,\pi)} \tag{5.3.4}$$

$$= \frac{\binom{s}{r} p^r (1-p)^{s-r} [1 - (1-\pi)^r]}{1 - (1 - p\pi)^s}$$

The extreme cases are truncate selection ($\pi = 1$) and single selection ($\pi \to 0$). The latter gives an interesting result. Since $\lim_{\theta \to 0} (1 - \theta)^n = 1 - n\theta$, equation 5.3.4 reduces for single selection to

$$P(r \mid a > 0; s,p,0) = \binom{s-1}{r-1} p^{r-1} (1-p)^{s-r}. \tag{5.3.5}$$

Thus single selection is equivalent to complete selection of sibs of the proband. Unfortunately this result does not generalize to $\pi > 0$.

The assumption that π is constant is only an approximation which must be tested. Information comes largely from the distribution of a probands among r affected in ascertained families,

$$P(a \mid a > 0; r,\pi) = \frac{\binom{r}{a} \pi^a (1-\pi)^{r-a}}{1 - (1-\pi)^r}. \tag{5.3.6}$$

If most cases are isolated, no critical test of the constancy of π can be made.

Under certain conditions, the distribution of ascertainments among probands can yield information about π even from isolated cases. For example, if there are many independent sources of ascertainment (such as physicians, hospitals, birth and death certificates, and patient associations), with z_i the probability of ascertainment from the ith source, the probability that a proband have t ascertainments is the truncated Poisson distribution,

$$P(t \mid t > 0) = \frac{m^t e^{-m}}{t!(1 - e^{-m})} \tag{5.3.7}$$

where $\qquad \pi = 1 - e^{-m}, \qquad m > 0$

and $\qquad m = \Sigma\, z_i$

The distribution of t gives most of the information about π, since both isolated and familial cases can be used. Typically the distribution of probands among affected gives less than 1/4 as much information as the distribution of ascertainments per proband. However, estimates on the hypothesis of independent ascertainment and uniform ascertainment probability may be significantly heterogeneous. The ascertainment distribution will tend to underestimate π if ascertainment by one source makes another less likely, as one physician's referral tends to preclude another. On the other hand, the distribution of probands on the assumption of a constant ascertainment probability may overestimate π because two or more siblings may be examined or counted as probands, even though only one of them would have submitted to examination independently. Pooling the two independent sources of information gives an intermediate value that is closer to the lower estimate. Experience suggests that the pooled estimate is better than either one separately.

Unless the ascertainment model is clearly understood, ascertainments will not be defined appropriately. Multiple referrals by the same physician are usually not independent. For example, if a proband was ascertained through a hospital record, the referring and consulting physician should not be counted as separate sources of ascertainment unless they also report the proband independently of the hospital record.

The distribution of ascertainments per proband can take other forms. For example, there may be a small number of independent sources of ascertainment, the ith with ascertainment probability z_i, and $\pi = 1 - \prod_i (1 - z_i)$. The assumption of independence may be removed by taking a gamma distribution of mean number of ascertainments or a beta distribution of source-specific ascertainment probability. In one of these ways an acceptable estimate of π can be obtained, providing care is taken in defining and recording ascertainments.

5.4 Sporadic Cases

Even if many trait-bearers are due to simple genetic mechanisms and therefore occur in high-risk families, it is the rule rather than the exception for some cases to be *sporadic* due to mutations, phenocopies, technical errors, extramarital conceptions, rare instances of heterozygous expression of a recessive gene, chromosomal nondisjunction, polygenic complexes, etc. These sporadics, of different etiology from the high-risk cases, must be distinguished from *chance-isolated* cases whose siblings, although normal, have the same *a priori* risk, p, of being affected. Sometimes the distinction between sporadic and chance-isolated cases can be made phenotypically, for example by the use of a discriminant function between isolated and familial cases. More commonly, a phenotypic distinction is difficult or impractical, but the proportion of sporadic cases can be determined.

The first and most general method of estimation uses the distribution of isolated and familial cases. If x is the proportion of cases that are sporadic among all cases in the population, then the probability that an isolated case with $s - 1$ normal siblings be sporadic is $x/[x + (1 - x)(1 - p)^{s-1}]$, where p is the segregation frequency in high-risk sibships. A family with an isolated case, either sporadic or chance-isolated, is called *simplex*, and a family with two or more cases is called *multiplex*.

The concept of sporadic cases is based on the assumption that families may be divided into low-risk and high-risk categories. If the low risk is sufficiently small, nearly all cases in low-risk families are sporadic and nearly all multiplex families are high-risk. Assuming constancy of p and the ascertainment probability π, the distribution of r affected among s children under incomplete selection is

$$P(r \mid a > 0; x,s,p,\pi) = \begin{cases} \dfrac{sp\pi[x + (1 - x)(1 - p)^{s-1}]}{xsp\pi + (1 - x)[1 - (1 - p\pi)^s]}, & r = 1 \\[3mm] \dfrac{\binom{s}{r}p^r(1 - p)^{s-r}[1 - (1 - \pi)^r]}{xsp\pi + (1 - x)[1 - (1 - p\pi)^s]}, & r > 1 \end{cases} \qquad (5.4.1)$$

If x is sufficiently large and if interest is focussed on high-risk families, an efficient sampling scheme is to restrict attention to multiplex families. Distributions have been derived under this mode of selection, or conditional on two or more probands, or with at least one affected of a particular sex.

Segregation analysis in terms of x is strengthened if there is some quantitative difference between sporadic and high-risk cases. Then if μ_F is the mean of familial cases, μ_S the mean of sporadic cases, and y the proportion of isolated probands that are sporadic, the expected mean of isolated cases is

$$\mu_I = y\mu_S + (1 - y)\mu_F \tag{5.4.2}$$

or

$$y = \frac{\mu_F - \mu_I}{\mu_F - \mu_S}$$

Weak use of this principle consists in testing whether the means of isolated and familial cases are significantly different. Strong use requires that μ_S be known. For example, if familial cases are due to rare recessive genes, their mean inbreeding μ_F will exceed the mean in the general population, which we may take as characteristic of sporadic cases. Then if the number of isolated probands is n and the number of familial probands is N, we may estimate x as

$$x = yn/(N + n) \tag{5.4.3}$$

In the literature terms like sporadic, isolated, and familial are often used ambiguously. *Sporadic* is a subset of *isolated*, not a synonym. *Isolated* or *familial* without qualification applies to a sibship, and is not the same as isolated or familial *in the pedigree*. Terminological confusion is partly responsible for conflicting claims about frequencies of sporadic cases.

5.5 The Mixed Model

During the first half of this century, segregation analysis in man was limited to constant p without sporadic cases. This model sufficed to confirm hundreds of mendelian traits. In 1958 the development of computers permitted extension to sporadic cases. This was the beginning of *complex segregation analysis*, which recognizes recurrence risks as variable among families of a given mating type. The next step allowed for continuous traits and for discrete ones under a polygenic model.

At this point genetic epidemiologists had Hobson's choice between the generalized single locus model, which neglects cultural inheritance and genes of small effect (polygenes), and the polygenic model which neglects cultural inheritance and genes of large effect (major loci). Since the various modes of inheritance are not mutually exclusive, it was essential to elaborate a mixed model that includes a major locus and polygenes simultaneously. Morton and MacLean developed the mixed model for nuclear families, allowing for environment common to sibs but confounding genetic and cultural inheritance from parent to child (table 5.5.1). Numerical analysis is so heavy that assortative mating, indices, unusual relationships, and a full treatment of cultural inheritance have not been incorporated from path analysis, which remains the best method to resolve genetic and cultural inheritance and should usually precede segregation analysis. Good agreement has been found between estimates of variance components by the two methods.

In practice only one locus with two alleles is recognized. Two or more loci with rare dominant alleles cannot be detected by segregation analysis, whereas duplicate rare recessives can be detected by comparison of matings with different numbers of affected parents. Therefore the great simplification implied by a single locus with two alleles (to which there is no practical alternative) has a tolerable cost. The distribution within each major locus genotype is assumed to be normal and homoscedastic.

The mixed model changes segregation analysis, for which classification of individuals as normal or affected is no longer necessary or desirable. Instead, the investigator seeks a continuous variable that satisfies the model, and accepts a discrete classification only if the search fails. Both a dichotomy and a trichotomy (normal, intermediate, affected) are in use under inclusive and exclusive definition of intermediacy (fig. 5.5.1). As far as practicable, definition of affection should wait until segregation analysis of a continuous trait has provided a justification and the optimal thresholds. When a major locus has been convincingly demonstrated on other grounds, such as bimodality for PTC sensitivity, segre-

Table 5.5.1: Parameters of the mixed model for pedigree analysis

Symbol	Definition
V	Phenotypic variance in population
U	Mean in population
t	Displacement between major locus homozygotes
d	Dominance at major locus
q	Gene frequency at major locus
H	Heritability in childhood
HZ	Adult heritability
x	Sporadic frequency due to mutation
m	Selection coefficient (sex linkage only)

In nuclear families H and HZ may be parametrized in terms of genetic heritability (H) and environment common to sibs (B), assuming no environmental transmission from parent to child. This does not extend to larger pedigrees.

gation analysis under the mixed model provides a test of whether dominance is complete and whether residual familial variation (polygenic or
environmental) is negligible. Even when a major locus is unlikely, segregation analysis has the advantage over path analysis that adequate
allowance is made for incomplete ascertainment.

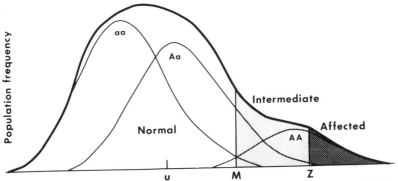

Fig. 5.5.1: The mixed model when the quantitative trait is trichotomized with affection excluded from intermediacy. M and Z are the two thresholds, and u is the mean.

Under the mixed model, segregation frequencies lose their role as
basic parameters and become functions of more fundamental variables,
which predict recurrence among as well as within mating types.

Segregation analysis is much more sensitive than path analysis to
departures from the distributional assumptions (table 5.5.2). Skewness
can simulate a nonexistent major locus, while elimination of skewness
can weaken evidence for a real locus. Reduction of a quantitative trait to
a trichotomy avoids distributional problems, but with considerable loss
of power. The ability of a polygenic model to fit data on a major locus is
remarkable. A large effect, low gene frequency, and dominance favor
resolution, but judicious use of power transformation, polychotomiza-
tion, selection through affected relatives, and tests of heterogeneity
among mating types may be required to detect or exclude a major locus.

Mutation at the major locus has been incorporated into the mixed
model. Other sporadic cases are not explicitly allowed for, since low
penetrance in the normal homozygote implies sporadic cases. Outliers
for quantitative traits present a special problem. If they are not part of
the normal distribution of small effects, they may simulate a rare recessive allele. This can be avoided by basing the power transform on a
truncated distribution, abnormal values being classified simply as *affected*, which is also appropriate if diagnosis leads to modification of the
quantitative trait. With this precaution, phenocopies should not simulate
a major locus.

Table 5.5.2 : Comparison of genetic and cultural heritability from path and segregation analysis

Trait	Path analysis		Segregation analysis	
	h^2	c^2	H	B
VLDL cholesterol	.46	.01	.33	.08
LDL cholesterol	.39	.12	.28	.14
HDL cholesterol	.31	.11	.38	.02
Triglyceride	.18	.08	.37	0
Cholesterol	.49	.04	.51	.06
Lipemia	.29	.06	.48	0
Uric acid	.27	.12	.34	0
Systolic BP	.24	.16	.28	.08
Diastolic BP	.19	.09	.24	.06

There is rough agreement. Because of covariance adjustment for the index, estimates of genetic heritability from segregation analysis (H) tend to be greater than estimates from path analysis (h^2). Conversely the estimate of cultural heritability from segregation analysis (B) tends to be smaller than from path analysis (c^2).

5.6 Dominance and Epistasis

The mixed model assumes that polygenes, major locus, and environmental effects are additive on the liability scale, with dominance limited to the major locus and without *epistasis* (interaction of nonallelic genes). The primary reason for these assumptions is that in nonexperimental material polygenic dominance and epistasis are not clearly resolvable from environment common to sibs. A secondary reason is that evidence for polygenic dominance and epistasis is weak in experimental organisms, since it comes from studies in which the environment was not effectively randomized within families and no effort was made to resolve major loci.

There is a long tradition of insistence on polygenic dominance and epistasis, rooted in the controversy between biometricians and mendelists at the beginning of the century. The mendelists won by showing that correlations of relatives could be predicted from the mendelian laws with any distribution of dominance and certain distributions of epistasis. However, neither then or subsequently was evidence presented that polygenic dominance and epistasis are important in nearly panmictic populations of higher organisms.

Several arguments have been raised against traditional emphasis on dominance and epistasis. (a) Detrimental genes have been shown to approach additivity as the degree of homozygous impairment decreases. (b)

Even quantitative effects of major genes often approach additivity: for example, alleles distinguished by electrophoresis typically have additive effects on enzyme activity. (c) On mathematical grounds, small effects are expected to be nearly additive (Maclaurin's theorem). (d) Interactive effects that may be detectable in crosses of inbred lines are usually small within a randomly mating population at stable gene frequencies. (e) Neglect of dominance and epistasis for polygenes is therefore a plausible hypothesis that can be tested by observations on family resemblance or inbreeding effects. The latter evidence is discussed in chapter 7.

5.7 Mutation

In classical segregation analysis mutation enters through the proportion x of sporadic cases in normal × normal matings. If alternative causes of sporadic cases (phenocopies, and so on) are negligible, the frequency of mutants may be estimated for autosomal dominants (section 8.3) and sex-linked genes (section 8.4). In pedigrees the parameter x is extended to all probands, without regard to mating type. Under sex-linkage, an additional parameter is required to reflect possible differences in mutation rate between sperm and egg. It is convenient to express this in terms of m, the selection coefficient against affected males. Transmission matrices from an individual to a descendant have been derived as a function of x for the autosomal case and of x and m for sex-linkage.

5.8 Pointers

There must be information in pedigrees that is lost by partition into nuclear families. We seek to extend segregation analysis beyond nuclear families under plausible assumptions about ascertainment. This leads to the definition of a class of phenotypically selected individuals called pointers, who may be either probands or secondary cases. A *pointer* is defined as a relative of extreme phenotype (affected if the data are qualitative or with X greater than some preassigned threshold, if the data are quantitative), through whom the family was selected. A pointer is outside the nuclear family pointed to. Pointers should not be assigned to a family that was selected through parents or children without consideration of family history. Otherwise at least one pointer is necessary, but may not be sufficient to assure ascertainment.

Each component of a nuclear family (father, mother, or the set of children) may have one pointer, taken to be the closest eligible relative if two or more are eligible, a given pointer being assigned only to the closest component of the nuclear family. Information about a pointer includes its phenotype and relationship to the family member. Thus a nuclear family with up to three pointers becomes the unit of pedigree analysis.

To make the calculations manageable, some simplifying assumptions are necessary. Relatives in the chain of kinship from pointer to parent are treated as of unknown phenotype, and other affected relatives are ignored. Then probabilities for the major locus genotype of the pointer conditional on the genotype of the closest family component are given by transmission matrices with mutation. Likelihood of the set of children's phenotypes is usually taken conditional on parent and pointer phenotypes.

The advantages of this approach are obvious. Likelihood calculations under the mixed model with mutation are manageable, and allowance for ascertainment bias is rigorous. However, three disadvantages must be considered. First, there must be some loss of information when pedigrees are partitioned, but nothing much can be said about this since no analysis of intact pedigrees under incomplete ascertainment has yet been validated, iterative calculations under the mixed model are only feasible with small pedigree structures, and calculations under simpler models are inadequate to resolve a major locus against a background of polygenic or cultural inheritance. Secondly, although an individual appears only once as a child in a conditional probability, he may also enter as a parent or pointer into the prior probability of another sibship. This introduces no bias but a slight dependence among nuclear families, the effect of which has never been observed in segregation analysis and is presumably negligible. This may be tested by grouping pedigrees at random and testing heterogeneity of segregation parameters among the partitions. If pointer logic is valid, this test will have no more than its nominal type I error. Thirdly, if pedigrees are sampled at random, without regard to affection, the parental phenotypes give information that is not extracted by likelihood conditional on parental phenotypes. We may optionally calculate joint instead of conditional likelihood, but joint likelihoods on the assumption of random sampling are invalid and seriously misleading under incomplete ascertainment and are sensitive to distributional assumptions. Pedigree calculations are feasible for a single pedigree without iteration, and so there is no reason to partition pedigrees in genetic counseling, which uses the estimates from segregation analysis. Partition into nuclear families gives specific recurrence risks and many useful tests of goodness of fit. Heterogeneity among mating types with 0, 1, or 2 parents affected suggests that there is more than one major locus and provides an estimate of the number involved. Heterogeneity among types of ascertainment and pointer configurations suggests that ascertainment has been incorrectly interpreted or the simplifying assumptions are invalid. Conversely, a good fit by these tests increases confidence in the conclusions of segregation analysis with pointers.

5.9 General Pedigrees

Two difficulties lie in the way of extending segregation analysis beyond pointers. First, the calculations become heavy under the mixed model. Secondly, the mode of ascertainment may be unknown or complex. These problems have been addressed in several ways, with limited success.

Computational difficulties arise when integration must be done numerically. If the likelihood is evaluated many times, computation is prohibitively slow. If the likelihood is evaluated only a few times, numerical integration is unreliable. The best compromise is Gauss-Hermite quadrature with appropriate scaling, which gives good accuracy in a small number of evaluations. The success of this approach may be determined by showing that the likelihood is stable as the number of evaluates increases. One alternative approach is to sample the possible genotypes for a pedigree. The computations are intolerable if the sample is large, unreliable otherwise, especially for a rare gene.

Another alternative is to abandon the mixed model. Elston and Stewart introduced transmission frequencies with mendelian expectations 0, 1/2, 1 for a major locus or polygenes. If significant departure from these values is found, they infer, but cannot estimate, cultural inheritance. However, other departures from the model with respect to ascertainment, age of onset, distribution of phenotype, and other assumptions can be reflected in spurious distortion of transmission frequencies. Cultural inheritance and any of these disturbances could coexist with a major locus and/or polygenes, which are not discriminated. Therefore, nothing much can be inferred from estimates of transmission frequencies. Abandonment of the mixed model destroys the rationale for segregation analysis, which is to test for a major locus without neglecting continuous variation.

Pedigree calculations are difficult, even under mendelian models, if there is inbreeding. This problem is minimized with nuclear families. Although troublesome in isolates, inbreeding is too rare in developed countries to pose any great practical problem.

Computational difficulty is a less serious problem than ascertainment bias in pedigrees. The simplest case is single selection, with pedigree form independent of content. The second condition means that pedigrees are extended at random or sequentially on the basis of phenotypes already sampled, but never selectively because of family history; that affection does not alter fertility; and that once a decision to extend the sample is made, the material will be included in analysis regardless of its content. Under these stringent assumptions, the probability of the pedigree as $\pi \to 0$ is analogous to equation 5.3.5, or

$$P(\text{pedigree} \mid \text{ascertainment}) = \frac{P(\text{pedigree})}{P(\text{proband affected})}$$

This result due to Cannings and associates (1979) does not extend to $\pi > 0$ or any other violation of the assumptions. Since any serious study in genetic epidemiology achieves a moderate to high ascertainment probability, the problem of incomplete selection in pedigrees remains unsolved. Nuclear families with pointers are a practical alternative.

5.10 Questions

1. *From equation 5.2.2, write the probability θ that an affected parent be a phenocopy if there are s children beyond the age of manifestation, all normal.*

$$\theta = \frac{h}{h + (1 - h)(1 - p)^s}$$

2. *In the above example, what is the mean risk for a subsequent child?*

$(1 - \theta)p$

3. *Generalize equation 5.2.2 to double segregation.*

$$P(r; h,y,s,p) = \begin{cases} h + (1 - h - y)(1 - p)^s, & r = 0 \\ (1 - h - y)\binom{s}{r}p^r(1 - p)^{s-r}, & 0 < r < s \\ y + (1 - h - y)p^s, & r = s \end{cases}$$

4. *Apply the above model to deaf-mute × deaf-mute matings, assuming phenocopies, at least one dominant gene, and more than one recessive gene.*

h = probability that at least one parent is a phenocopy.

y = probability that affection in both parents is due to homozygosis for the same recessive gene, so that the mating is $aa \times aa$.

p = backcross segregation frequency for dominant gene.

5. *Show that equation 5.3.4 obeys Bayes theorem.*

   A = segregation

   B = ascertainment

   $P(A|B) = P(A) \cdot P(B|A)/P(B)$

6. *Assume two independent sources of ascertainment with probabilities p_1, p_2. What is the probability of ascertainment by source 1 only, given ascertainment by at least one source?*

$$\frac{p_1(1 - p_2)}{1 - (1 - p_1)(1 - p_2)}$$

7. *What is the effect of assuming single selection when $\pi > 0$?*

   The segregation frequency p is underestimated, and the proportion x of sporadic cases is overestimated.

8. *The proportion x of sporadic cases for limb-girdle muscular dystrophy has been estimated as .349 ± .058 by segregation analysis and .500 ± .075 from the inbreeding coefficients of isolated and familial cases. Test these estimates for consistency and give the efficient pooled estimate.*

   Information: $\frac{1}{.058^2} = 293$, $\frac{1}{.075^2} = 176$

   pooled estimate = $[.349(293) + .500(176)]/(308 + 173) = .41$

   $X_1^2 = (.349 - .41)^2(293) + (.50 - .41)^2(176) = 2.5$,

   Not significant (NS)

9. *Discuss the stability of the above value of x.*

   If due to rare expression in heterozygotes of usually recessive genes, it is expected to be stable. If due to phenocopies (eg, misdiagnosis of polymyositis and neurological disease as limb-girdle muscular dystrophy) it will decrease with improved diagnosis. Resolution of this uncertainty, which is serious for genetic counseling, requires new studies in genetic epidemiology.

10. *An investigator is analyzing a quantitative trait where a major locus has not been demonstrated. Should he transform to eliminate skewness? Why?*

    Yes, to protect against misinterpreting skewness as a major locus.

11. *A major locus for a quantitative trait has been supported by segregation analysis. For genetic counseling and linkage, should the trait be transformed to eliminate skewness in the general population? Why?*

    No. To give the most accurate description of the phenotype distribution skewness should be abolished *within* major locus genotype, not overall.

12. *Transmission frequencies for a certain trait are found to be significantly nonmendelian. A. What are the possible explanations? B. How could they be discriminated?*

    A. Cultural inheritance, misspecification of major locus effects, failure of distributional assumptions, etiologic heterogeneity, and ascertainment bias.
    B. By analysis under the mixed model.

13. *Transmission frequencies for a certain trait are significantly unequal, but not significantly different from mendelian expectation. A. What are the likely explanations? B. How could they be discriminated?*

    A. Major locus or polygenic inheritance.
    B. By analysis under the mixed model.

14. *Transmission frequencies for a certain trait are not significantly unequal. What can you conclude?*

    The sample has low power to discriminate modes of inheritance.

15. *Suppose the variance of sibs of probands is greater than the variance of sibs in unselected families. Discuss.*

This is consistent with, but not critical evidence for a major locus. Other possibilities include correlation between mean and variance, either primary or secondary to an effect on both of age, maternal phenotype, or other familial factor.

16. *Besides the above variance ratio, other criteria outside segregation analysis have been suggested to discriminate a major locus. What do such methods have in common?*

Lack of specificity in tests of significance, inability to estimate parameters of the hypothetical major locus, and failure to provide tests of etiological heterogeneity.

17. *What are the advantages and disadvantages of joint likelihood of parents and children under incomplete selection?*

Information can be extracted from the distribution of parental phenotypes, but this information is invalidated if there is assortative mating or an effect of phenotype on fertility, or if any parent is a proband. Heterogeneity among mating types cannot be studied, and sensitivity to distributional assumptions is increased.

18. *Does rejection of a major locus for one trait extend to a correlated trait?*

No. Greater displacement may give significant evidence for a locus that is not detected when the displacement is small. Displacement may be amplified by repeated or more precise measurement, or by determination of a correlated trait more closely related to primary action of the locus.

19. *Using the symbols of table 5.5.1, express the condition for a megaphenic effect.*

$t/\sqrt{V} > 1$ if $d \to 0$

$dt/\sqrt{V} > 1$ otherwise

20. *Marxist critics of genetic epidemiology have claimed that heritability is "useless" for segregation analysis that is the basis of genetic risks. Clarify.*

This is a rational opinion if and only if they can provide a different method for segregation analysis which gives a demonstrably better fit than the mixed model. No such attempt has been made.

5.11 Bibliography

Cannings C, Thompson EA, Skolnick M: Extension of pedigree analysis to include assortative mating and linear models. In: The Genetic Analysis of Common Diseases: Application to Predictive Factors in Coronary Heart Disease. Edited by Sing CF, Skolnick M. Alan R. Liss, New York, 1979

Elandt-Johnson RC: Probability Models and Statistical Methods in Genetics. John Wiley and Sons, New York, 1971

Elston RC, Stewart J: A general model for the genetic analysis of pedigree data. Hum Hered 21: 523–542, 1971

Lalouel JM: Probability calculations in pedigrees under complex modes of inheritance. Hum Hered 30: 320–323, 1980

Morton NE: Genetic tests under incomplete ascertainment. Am J Hum Genet 11: 1–16, 1959

Morton NE: Segregation analysis. In: Computer Applications in Genetics. Edited by Morton NE. University of Hawaii Press, Honolulu 1969, pp 129–139

Morton NE, Maclean CJ: Analysis of family resemblance. III. Complex segregation of quantitative traits. Am J Hum Genet 26: 489–503, 1974

Morton NE, Yasuda N: Transition matrices with mutation. Am J Hum Genet 32: 202–211, 1980

6. Association

Affection status (as normal *vs* affected) or a quantitative trait underlying affection may be associated in families or in the population with another phenotype, either genetic or not. There are many causes for such associations, of which some of the most important are allelism, linkage, pleiotropy, interaction, and nonrandom distribution of genes and environment.

6.1 Allelism

The frequencies of multiple alleles are negatively correlated. The frequencies of factors determined by a multiple allelic system may be correlated, either negatively or positively. Table 6.1.1 shows the number of observations on two factors and the common measure and test of significance of their association. Many allelic factor pairs are so weakly associated that very large samples would be required for statistical significance. Of 55 correlations between allelic factors studied in northeastern Brazil, 11 were in the range from $-.05$ to $+.05$. To detect a correlation of $\pm .05$ at the .01 significance level requires $\chi^2 = 6.635$ and a sample size of $n = 6.635/(.05)^2 = 2654$. Thus the power to detect allelic factor association is far from complete even in large samples.

Information about association under the null hypothesis is

$$K = p_1(1 - p_1)p_2(1 - p_2)n \tag{6.1.1}$$

where p_1, p_2 are the phenotypic factor frequencies. This shows that the power to detect association is maximal with frequencies near .5. The information is greatly reduced if both factors are rare or very common. For example, if $p_1(1 - p_1) = p_2(1 - p_2) = .01$, the information per observation is only $1/625$ of its maximal value. In verification of this, of 25 allelic associations in northeastern Brazil involving a factor with $p(1 - p)$ less than .05, only 5 were significant at the .01 level. However, of 37 pairs with $p(1 - p)$ between .05 and .15, there were 29 significant associations.

Table 6.1.1: Factor associations

Factor 1	Factor 2		Total
	+	−	
+	a	b	$a + b$
−	c	d	$c + d$
Total	$a + c$	$b + d$	$a + b + c + d = n$

Product-moment measure of association:

$$r = \frac{ad - bc}{\sqrt{(a + b)(c + d)(a + c)(b + d)}} \doteq \sqrt{\frac{\chi^2}{n}}$$

Yates test of significance of association:

$$\chi_1^2 = \frac{(|ad - bc| - n/2)^2 n}{(a + b)(c + d)(a + c)(b + d)}$$

Of 48 pairs with $p(1 - p)$ greater than .15, there were 46 significant associations. At the .01 level, 27 percent of the observed correlations of factors within systems would be significant with a sample size of $n = 100$, 73 percent with $n = 1000$, and 85 percent with $n = 10,000$.

We see that although many associations of allelic factors can be detected with moderate sample sizes, others cannot. The investigator is fortunate if he can guess the system to which a new factor belongs through a test of association, but he should not give much weight to a nonsignificant result. Complete absence of one phenotypic class is more helpful, for example Kell factors among $k_0 k_0$ subjects and Hp 1F-1S phenotypes among Hp 2-1 persons. However, only a linkage test provides powerful and unambiguous corroboration that a factor does, or does not, belong to a given system.

6.2 Linkage

The organization of genes into linear structures, the chromosomes, restricts segregation of pairs of systems. The tendency of two genes on the same chromosome to be inherited together is called *linkage*. Two systems located close together are separated by recombination less often than two systems far apart, and this fact makes it possible to arrange the various systems within a chromosome in a linear map according to their recombination frequencies. Two genes are said to be in *coupling* if they entered a zygote in the same gamete, and in *repulsion* otherwise. Coupling of linked genes implies presence on the same chromosome of a homologous pair. If θ is the frequency of recombination between two systems G and T, there will be association within pedigrees when $\theta < 1/2$ (table 6.2.1). However, association will not be found in the population unless the frequencies of coupling and repulsion are unequal. We saw in section 3.8 that the approach to equality is rapid unless θ is small. Therefore population admixture virtually destroys association, except for closely linked genes.

Detection of linkage encounters two problems. First, the number of critical families is often small enough so that large-sample tests are unreliable. Secondly, the chance that two random genes be linked is small (about .054), and so a large fraction of weakly "significant" linkages is spurious. Both difficulties are removed by the lod score method, which uses

$$z_i(\theta) = \log_{10} [P_i(\theta)/P_i(1/2)] \tag{6.2.1}$$

as the lod score for θ in the ith pedigree, where $P_i(1/2)$ is the probability under the null hypothesis that $\theta = 1/2$ (no linkage). Then, on this null hypothesis, the inequality

$$P\{z_i(\theta) > \log A \,|\, H_0\} < 1/A \tag{6.2.2}$$

is exact even in small samples, and provides strong evidence for linkage if A is sufficiently large. It is customary to take $A = 1000$ (ie, $\log_{10} A = 3$), which assures that most significant scores are due to linkage. In general recombination between linked loci is less in males, and so we must distinguish θ_m in males and θ_f in females.

Table 6.2.1: Gametic output of a double heterozygote $GgTt$

	GT	Gt	gT	gt
GT coupling	$\dfrac{1-\theta}{2}$	$\dfrac{\theta}{2}$	$\dfrac{\theta}{2}$	$\dfrac{1-\theta}{2}$
GT repulsion	$\dfrac{\theta}{2}$	$\dfrac{1-\theta}{2}$	$\dfrac{1-\theta}{2}$	$\dfrac{\theta}{2}$
Observed	a	b	c	d

Let λ be the probability the G and T are in coupling
$P(a, b, c, d) = \lambda(1 - \theta)^{a+d}\theta^{b+c} + (1 - \lambda)\theta^{a+d}(1 - \theta)^{b+c}$

The linkage map includes not only recombinational evidence, but also the distribution of chiasmata in spermatogenesis and physical assignments by cytogenetics and somatic cell genetics. Methods have been developed to transform one map into another, and to test one or more markers for linkage to a chromosome.

Linkage can be a powerful tool to resolve two groups of loci, one linked to a marker. For example, two dominant genes for elliptocytosis are known, one (*ELI*) close to the *RH* locus on chromosome 1. With large pedigrees and tight linkage this heterogeneity is apparent by inspection of lod scores (fig. 6.2.1), reinforced by the large-sample test of heterogeneity among n pedigrees,

$$\chi^2_{n-1} = 4.605\,[\Sigma z_i(\hat{\theta}) - \hat{Z}] \tag{6.2.3}$$

where $z_i(\hat{\theta})$ is the maximum lod in the ith pedigree, $\hat{Z}$ is the maximum overall, and $4.605 = 2\ell n 10$. A more powerful test is to write the probability of the ith pedigree as

$$P_i(\theta,m) = mP_i(\theta) + (1-m)P_i(1/2) \tag{6.2.4}$$

where m is the proportion of pedigrees with linkage, estimating θ and m simultaneously. Large pedigrees, dominance, complete penetrance, and close linkage favor resolution of linkage heterogeneity.

Tests of linkage heterogeneity can be made simpler and more efficient by expressing recombination in one sex as a function of recombination in the other. Let w_m, w_f be the map distances in males and females, respectively, with $k = w_f/w_m$. The average value of k is about 1.8, increasing around the centromere and decreasing distally. If k is known approximately for a given chromosome segment, we may take it as constant and tabulate lod scores in terms of a single variable, θ_m, without making the arbitrary and generally incorrect assumption that $\theta_f = \theta_m$. This simplification is especially useful for small pedigrees and low penetrance. The mapping function that best fits human data is

$$w = \{p(2p-1)(1-4p)\ell n(1-2\theta) + 16p(p-1)(2p-1)\tan^{-1}(2\theta)$$

$$+ 2p(1-p)(8p+2)\tanh^{-1}(2\theta) + 6(1-p)(1-2p)(1-4p)\theta\}/6 \tag{6.2.5}$$

with $p = .351$. Since map distance is defined as number of crossovers, it has the property of additivity, which is required to construct a genetic map.

Tight linkage often leads to *gametic disequilibrium*, defined as a

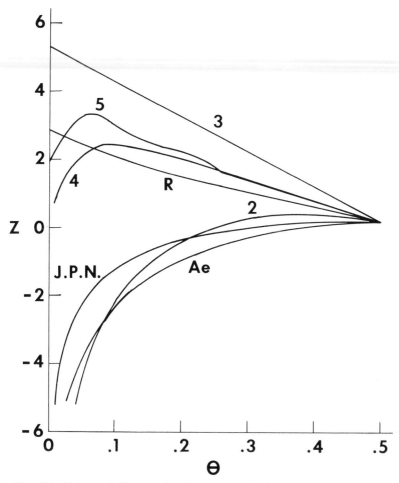

Fig. 6.2.1: Linkage of elliptocytosis with *RH* in 4 of 7 families. At a recombination value of $\theta = .05$, the lod score Z is greater than 2 in families 3,4,5, and R (linkage) and less than 2 in families 2, Ae, and JPN (no linkage).

coupling frequency different from 1/2. In the most extreme case, this induces an association in populations which may be difficult to distinguish from multiple allelism. Critical evidence for gametic disequilibrium requires observation of at least one unambiguous recombinant in pedigrees.

6.3 Pleiotropy

A particular gene may have multiple phenotypic effects. This is called *pleiotropy*. Distinction from gametic disequilibrium can be difficult. Some of the most interesting cases of pleiotropy are provided by genes that affect cell surfaces and therefore regulate antibody response and transport through the cell membrane.

In the *ABO* system, gene *A* is weakly but consistently associated with stomach cancer and other neoplasms of the digestive system. These cancers tend to produce the Forssman antigen, which crossreacts with anti-A. Individuals of types A and AB (who cannot produce anti-A) are therefore less able to recognize and destroy cells marked with the Forssman antigen. Gene *O* is associated with duodenal ulcer. Susceptibility to rheumatic carditis is greatest in persons who are of group A, B, or AB and are unable to secrete, in water-soluble form, the specific antigens whose presence on cell membranes, in alcohol-soluble form, is controlled by the *ABO* system. The genotype *O/O SE/-* secretes H substance, which is apparently protective against Streptococcus bacilli. Severe infection by this pathogen causes an increased risk of rheumatic carditis.

The HLA system on chromosome 6 is an exceedingly complicated cluster of loci that control recognition of foreign antigens, interspersed with other loci that control levels of certain complement components and enzymes. A homologous system is found in higher vertebrates. Some disease associations of the *HLA* system are as weak as for the *ABO* blood groups, but others are much stronger.

The strength of disease association is best measured by relative risk, which is invariant with respect to disease frequency. If the odds for disease are $p/(1 - p)$ in low-risk genotypes and $q/(1 - q)$ in high-risk genotypes, the *relative risk* is

$$x = \frac{p(1 - q)}{(1 - p)q} \qquad (6.3.1)$$

In table 6.3.1, this is expressed in terms of the observed frequencies (a, b, c, d). Since the estimate of relative risk is meaningless if any of the frequencies is zero, a small bias correction is included. The maximum likelihood test of association is unbiased and may be used even in small samples. All estimates of relative risk require that the marginal totals $(a + b, c + d, a + c, b + d)$ be nonzero.

Disease associations are ordinarily studied by determining the frequency of the genetic factor in a sample of cases with the disease, and in another sample of controls without the disease. In practice, the control

Table 6.3.1 : Disease associations

Genetic factor	Disease	
	+	−
+	a	b
−	c	d

Relative risk

$x = (a + \tfrac{1}{2})(d + \tfrac{1}{2})/(b + \tfrac{1}{2})(c + \tfrac{1}{2})$

$y = \ln x$

$v = \text{Var } y = \dfrac{1}{a + 1} + \dfrac{1}{b + 1} + \dfrac{1}{c + 1} + \dfrac{1}{d + 1}$

Combined relative risk: $\bar{x} = e^{(\Sigma y/v)/\Sigma(1/v)}$

Combined test for association: $\chi_1^2 = (\Sigma y/v)^2/\Sigma(1/v)$

Test for heterogeneity among m samples: $\chi_{m-1}^2 = \Sigma(y^2/v) - \chi_1^2$

Maximum likelihood test of the null hypothesis $E(x) = 1$ in small samples

$u = (ad - bc)/n$

$k = (a + b)(c + d)(a + c)(b + d)/n^3$

Combined relative risk: $\bar{x} = \Sigma u/\Sigma k$

Combined test for association: $\chi_1^2 = (\Sigma u)^2/\Sigma k$

Test for heterogeneity among m samples: $\chi_{m-1}^2 = \Sigma(u^2/k) - \chi_1^2$

Fisher's F test: $F = (m - 1)\chi_1^2/\chi_{m-1}^2$, $df = 1, m - 1$

sample, which should be matched for race and other relevant factors, may be difficult to select. Usually relative risk increases with stringency of case selection, which varies among studies. Therefore heterogeneity in relative risk is often observed.

Linkage tests can be misleading when there is pleiotropy, since misspecification of the mode of inheritance tends to simulate recombination. Linkage analysis does not usually include polygenes, cultural inheritance, or etiological heterogeneity, and so inheritance will often be misspecified.

Disease associations may be explained in terms of gametic disequilibrium, which in a nearly panmictic population is improbable for recombination values greater than .01. A significantly larger value within racial groups, in conjunction with disease association, is indicative of pleiotropy with etiological heterogeneity or other misspecification of the mode of inheritance, and should not be accepted as evidence for loose linkage of a locus for disease susceptibility in genetic disequilibrium with the marker system. Properly interpreted, linkage analysis can provide a valuable test for homogeneity of disease association.

6.4 Interaction

A set of phenotypically interactive systems (epistasis) cannot be identified in man by current methods, unless each system has been individually characterized. For example, the *LE* system controls a water-soluble antigen which is adsorbed on cells (table 6.4.1). When tested in saliva, there is no interaction with the unlinked *SE* system that controls ABH secretion. However, when tested on red blood cells, the Lewis antigen types as Le(b) in ABH secretors and as Le(a) in nonsecretors. The *LE* and *SE* systems control enzymes that attach specific carbohydrates to the same mucopolysaccharide. These radicals behave independently in solution, but interact antigenically when constrained by attachment to a cellular membrane. Recognition of this hypersystem depended on antibodies specific for the *LE* and *ABO* systems. Attempts to understand inheritance of the Lewis factors on red blood cells were unsuccessful until the specific effect of ABH secretion had been recognized and attention was directed to the main effect of the *LE* system in secretions. This paradigm justifies omission of epistasis from segregation analysis.

6.5 Age, Sex, and Environmental Associations

The experimental geneticist observes each generation at a fixed age. This is usually not possible in genetic epidemiology. The "same" trait observed at different ages is really a series of correlated traits, acted on by different genes and environments. This is imperfectly described by conventional path analysis of family resemblance, but can be revealed by comparison of correlations at different ages. Parent-child correlations often increase with age of the child as the measured trait converges to its adult value. Resemblance of sibs often decreases with the difference in

Table 6.4.1: The Lewis-secretor hypersystem. There is interaction on red cells but not in saliva

Lewis genotype	Secretor genotype	Red cell phenotype	Saliva phenotype	
			ABH	Lewis
1/1 or 1/0	1/1 or 1/0	Le(b)	+	Le(a)
	0/0	Le(a)	0	Le(a)
0/0	1/1 or 1/0	Le(−)	+	Le(−)
	0/0	Le(−)	0	Le(−)

their ages, being greatest for twins and least for children with several intervening sibs (fig. 6.5.1). Resolution of genes and environment in temporal trends is not feasible unless the same individual is measured at different ages. Temporal trends are not evidence for gene-environment interaction, since the same trends might be observed in a completely inbred line.

Gene-environment interaction can legitimately be claimed only when both the relevant genes and environments have been identified. In principle this could be realized with polygenic genotypes in somatic cells, but in practice the study of gene-environment interaction has been limited to drug sensitivity of major locus genotypes. The antimalarial drug primaquine has no side effects in most individuals, but produces hemolytic anemia in males carrying *gpd* alleles at the sex-linked locus for glucose-6-phosphate dehydrogenase. The muscle relaxant suxamethonium is innocuous to most patients, but may produce severe respiratory difficulty in carriers of the rare alleles *S*, *F*, and *R* at the *E*1 locus for pseudocholinesterase. Idiosyncratic drug sensitivity is the subject of *pharmacogenetics*, a branch of genetic epidemiology.

Age, sex, and disease associations cannot be understood unless the genetic and environmental components are precisely specified. In genetic

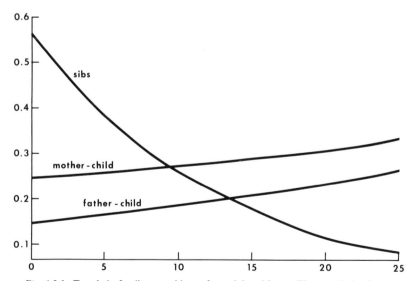

Fig. 6.5.1: Trends in family resemblance for weight with age. The correlation between sibs decreases with the difference in their ages. Familial resemblance for birth weight is largely determined by the maternal constitution and environment, but this maternal effect diminishes with age as the parent-child correlation increases.

epidemiology representation of gene-environment interaction as a variance component is rarely justified except under implausible and untested assumptions. Response to an environmental agent can be treated as a main effect by standardizing dosage. When feasible this is simpler, less controversial, and more informative than representation as a variance component due to gene-environment interaction, which in studies of family resemblance is confounded with errors of measurement in the unexplained residual variance.

6.6 Group Differences

When the genetic epidemiologist stratifies a population into groups, he hopes that differences among them are entirely genetic or entirely environmental. Otherwise, genotypes and environments are said to be confounded. In the absence of strict randomization of the environment, there is no infallible method to resolve environmental and genetic factors that differ between groups. Instead several methods permit cautious inferences.

If the difference is environmental, valid covariance adjustment and stratification should reduce its magnitude. When these techniques make the difference nonsignificant, the motivation to look for genetic factors is weakened. However, it may be impossible to prove validity if genetic and environmental factors are confounded. Table 6.6.1 shows how racial differences in academic test achievement diminish with covariance adjustment.

Table 6.6.1: Effect of covariance adjustment on mean test achievement of six racial groups

		Analyses		
Group	Before	1	2	3
Indian American	44.0	47.6	48.2	48.6
Mexican-American	42.0	45.5	46.4	47.4
Puerto-Rican	38.3	43.8	45.4	47.1
Negro	42.3	45.3	44.9	49.3
Oriental-American	49.3	50.5	50.6	51.1
White	53.0	51.7	51.6	50.5

For each analysis, overall mean is 50 and standard deviation is 10. Analysis 1 adjusted for socioeconomic status and family structure and stability; analysis 2 adjusted for home and family background; and analysis 3 adjusted for all variables.

Because covariance analysis and stratification assume that the independent variables are measured without error, residual differences cannot be interpreted. Path analysis allows us to test causal hypotheses about correlated factors, even when the causes are only estimated. Figure 6.6.1 gives a path analysis of the same data. The relative direct effect (RE) is the proportion due to direct causation relative to total effect, or

$$RE(ij) = \frac{p_{ij}}{p_{ij} + \sum_k p_{ik} p_{kj}} . \qquad (6.6.1)$$

In model 3, the only determinate model that fits the data, the RE for race on performance is .2, and so 80 percent of the difference between two racial groups is attributed to social class and would be expected to disappear if children were randomized among social classes, other factors remaining the same. The residual 20 percent is not explained, and may well be due to social pressures based on appearance and group membership, not collinear with class. No inference about heritability of this residual can be made from these data.

In this material the direct effect of school factors is nonsignificant. Apparently the public school system with its local school boards acts to perpetuate rather than to equalize group differences.

Results are considerably different in Hawaii, the only state without autonomous school districts. Three models are consistent with the evidence. Figure 6.6.2 shows the school characteristics have a small but significant effect on academic performance, while the direct racial effect is nonsignificant.

Although path analysis of individuals gives information about environmental causes, it relegates genetic factors to an unexplained residual, along with all environmental effects not collinear with estimated causes. To resolve a genetic component path analysis of intergroup matings or adoptions is required, with a racial index in addition to phenotype. Claims of genetic differences, not supported by a significant path from genotype to index, are unconvincing.

6.7 Maternal-fetal Incompatibility

Hemolytic disease of the newborn is due to maternal antibody against an antigen inherited from the father. Such *maternal-fetal incompatibility* occurs as a rare anomaly in many blood group systems, providing the most direct evidence that blood group polymorphism is not neutral. Maternal-fetal incompatibility is directed against heterozygotes, and

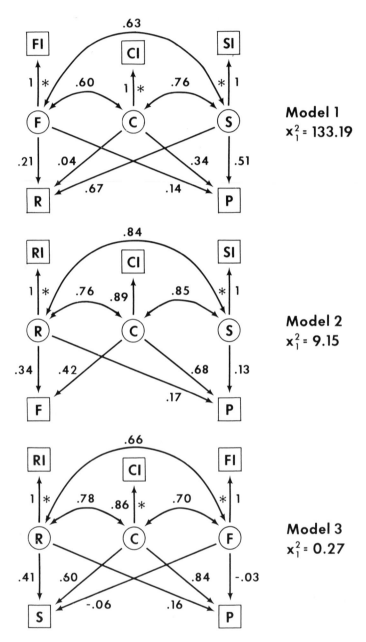

Fig. 6.6.1: Alternative 3-cause systems for academic performance. *R* = race, *C* = class, *S* = school, *F* = family structure, P = performance; causes followed by I denote the corresponding indices. Paths fixed by hypothesis are indicated by asterisks.

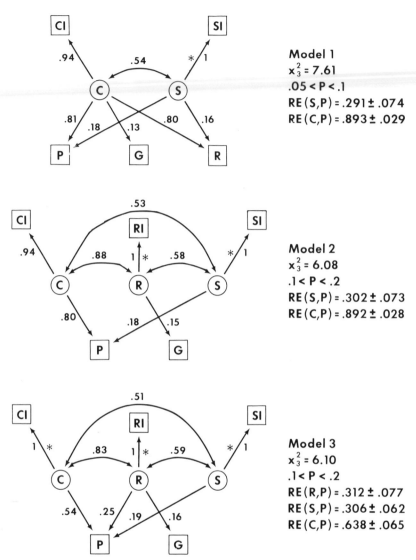

Fig. 6.6.2: Causal systems for academic performance in Hawaii. C = class, R = race, S = school, P = performance, G = grade score, I = index. Assumed parameters are indicated by asterisks.

therefore against the rarer allele, so that polymorphism must be maintained by other mechanisms.

In Caucasians the D factor of the *Rh* system has a gene frequency of about .6. Hemolytic disease in Rh-positive children with sensitized Rh-negative mothers was an important cause of stillbirth, neonatal death, and sometimes mental retardation until affected liveborn were treated by exchange transfusion. Subsequently it was found that Rh-hemolytic disease of the newborn could be prevented by administration of anti-Rh antibodies to mothers at risk.

The *ABO* system is the only one in which individuals lacking an antigen normally have the corresponding warm antibody. This makes compatibility for the *ABO* system essential in blood transfusion. Remarkably, ABO incompatibility does not often cause hemolytic disease of the newborn, although jaundice sufficient to require exchange transfusion occurs, and in some populations ABO incompatibility is an important cause of spontaneous abortion. IgM and IgA antibodies do not cross the placenta, and water-soluble antigen in secretor fetuses can bind IgG. The circumstances under which an ABO-incompatible fetus is at high risk remain obscure.

6.8 Questions

1. *How should segregation analysis under the mixed model be applied to linkage?*

   By fitting the parameters H and Z simultaneously if null values are not supported. Linkage analysis does not allow for these sources of variation, which cannot simulate linkage. Overestimation of the displacement t should be avoided, since it tends to give an overestimate of recombination.

2. *From figure 6.6.1, estimate what difference would persist if social class were randomized between two races which differ in mean IQ by 15 points?*

   $.2 \times 15 = 3$ IQ points.

3. *Two systems G and T have recombination frequency 1/2. Using equation A.1.3, what is the probability that a mating $GT/gt \times ggtt$ have no gg and/or no tt child among s children?*

   A_1 = no *gg* child

   A_2 = no *tt* child

   $A_1 A_2$ = no *gg* or *tt* child = all *Gg Tt*

$$P(A_1) = P(A_2) = (1/2)^s$$

$$P(A_1 A_2) = (1/4)^s$$

$$P(A_1 \cup A_2) = (1/2)^{s-1} - (1/4)^s$$

4. *Generalize problem 3 for arbitrary recombination fraction θ.*

$$P(A_1 \cup A_2) = (1/2)^{s-1} - [(1 - \theta)/2]^s$$

5. *Among families of size s having at least one gg and at least one tt child, what is the frequency of families with only ggtt children?*

$$\frac{(1 - \theta)^s}{2^s - 2 + (1 - \theta)^s}$$

6. *Show that incomplete ascertainment has an effect on linkage tests only if there is selection for both loci.*

Suppose there is selection on the main (G) locus but not on the test (T) locus. Then:

$$P(G,T) = P(G)P(T \mid G)$$

But $P(G)$ is not a function of θ, and $P(T \mid G)$ does not depend on how G was selected.

7. *Criticize the term "relative risk".*

It strictly denotes relative odds, which converge to relative risk as the disease becomes rare.

6.9 Bibliography

Conneally PM, Rivas ML: Linkage analysis in man. In: Advances in Human Genetics. Vol 10. Edited by Harris H, Hirschhorn K. Plenum Press, New York, 1980, pp 209–266

Human Gene Mapping 5: Fifth International Workshop on Human Gene Mapping. Birth Defects: Original Article Series XV, 11, The National Foundation, New York, 1979; also in Cytogenetics and Cell Genetics 25: 1–4, 1979

Keats BJB, Morton NE, Rao DC, Williams W: A Source Book for Linkage in Man. Johns Hopkins University Press, Baltimore, 1979

Mi MP, Morton NE: Blood factor association. Vox Sang 11: 434–499, 1966

Rao DC, Morton NE, Elston RC, Yee S: Causal analysis of academic performance. Behav Genet 7: 147–159, 1977

Svejgaard A, Hauge M, Jersild C, Platz P, Ryder LP, Nielsen LS, Thomsen M: The HLA system—An introductory survey. In: Monographs in Human Genetics. Edited by Beckman L, Hauge M., S. Karger, Basel, 1975

7. Population Structure

Population structure comprises all factors that determine mating frequencies. As defined, it poses a number of questions to the historian, sociologist, demographer, and other student of human behavior. Only two aspects of the problem are relevant to genetic epidemiology: the effect of population structure on gene frequencies and the relation between gene and genotype frequencies. This continues the classical concern of epidemiology with clustering of disease in time and space.

7.1 Kinship

Two individuals are said to be biological *relatives* if one is an ancestor of the other, or they have one or more common ancestors. Children of the same parents are called *sibs or siblings*, words that conveniently make no distinction between brothers or sisters. The children of sibs are first cousins, and their grandchildren are second cousins. Individuals with one parent in common are half sibs, and their children are half first cousins. The child of one's first cousin is called a first cousin once removed, and the grandchild of one's first cousin is a first cousin twice removed (fig. 7.1.1). Roman and canonical terms used in many countries are compared with their English equivalents in table 7.1.1.

The fundamental concept for studying resemblance between relatives is called identity by descent. Two genes are *identical by descent* if they are derived without mutation by transmission along a common path from the same gene in a common ancestor. *The (coefficient of) kinship ϕ_{IJ}* of individuals I and J is the probability that a randomly chosen gene from one be identical by descent with a randomly chosen allele from the other. If the ith genetic path between I and J has m_i lines, mutation being ignored,

$$\phi_{IJ} = \sum_{i=1}^{t} (\tfrac{1}{2})^{m_i + 1}, \qquad m = 0, 1, \ldots \tag{7.1.1}$$

This formula permits loops through a common ancestor to a more remote common ancestor. *Degree of kinship* k is defined by $\phi = (\tfrac{1}{2})^{k+1}$,

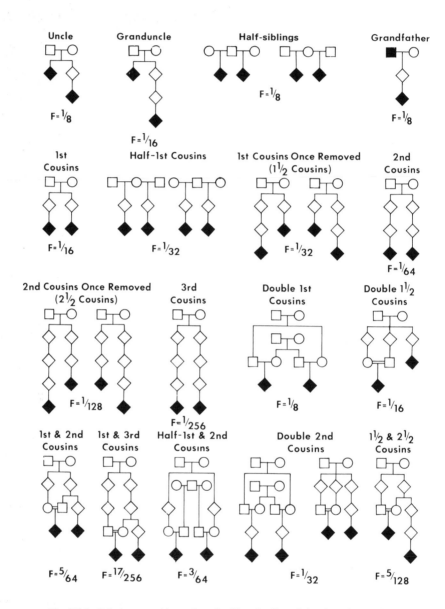

Fig. 7.1.1: Relatives outside nuclear families. Studies of the Atomic Bomb Casualty Commission introduced by translation from Japanese terms like "1½ cousin" as a succinct synonym of "first cousin once removed." □ Male; ○ female; ◇ male or female; ◆◆ defined relatives; = consanguineous marriage.

Table 7.1.1: A comparison of terms for relationship

English	Roman	Canonical	Kinship ϕ
Sibs	2	I/I	1/4
Uncle-Niece	3	II/I	1/8
First cousins	4 (cousins of first degree)	II/II	1/16
First cousins once removed	5 (cousins of second degree)	III/II	1/32
Second cousins	6 (cousins of third degree)	III/III	1/64
First cousins twice removed	6 (cousins of third degree)	IV/II	1/64

and so any value of kinship in the range $(\frac{1}{2})^{k+1.5}$ to $(\frac{1}{2})^{k+.5}$ is assigned degree k (table 7.1.2).

Kinship is by far the most useful measure of genetic similarity. It can be decomposed into identity coefficients, of which there are six in general, but only three if I and J are themselves not inbred (ie, the parents of I are not related, nor are the parents of J). Then I and J are called *regular* relatives. If C_i is the probability that two regular relatives have i genes identical by descent (i = 0, 1, 2), then

$$C_2 = \sum_{i \neq j} (\tfrac{1}{2})^{m_i + m_j - 2} \tag{7.1.2}$$

$$C_1 = \sum_{i \neq j} (\tfrac{1}{2})^{m_i - 1}[1 - (\tfrac{1}{2})^{m_j - 1}]$$

$$C_0 = 1 - C_1 - C_2$$

$$\phi_{IJ} = C_2/2 + C_1/4.$$

We say that I and J are *bilineal* relatives if there are at least two disjoint genetic paths between them, and *unilineal* otherwise. Clearly $C_2 = 0$ for unilineal relatives. Bilineal relatives of a rare recessive homozygote are at appreciable risk for affection. Sibs and monozygous twins are the only common type of bilineal relatives (table 7.1.3). Since dominance deviations are expressed through C_2, sibs are genetically more similar than parent-offspring pairs which have the same kinship. (Sibs also tend to have more similar environments, which makes estimate of the dominance variance impractical).

Following Cotterman, it is easy to derive the probability that an individual have a particular genotype, given the genotype of a relative.

Table 7.1.2: Kinship ϕ and degree of kinship k

Degree (k)	Typical $(\tfrac{1}{2})^{k+1}$	Range of ϕ $(\tfrac{1}{2})^{k+1.5} - (\tfrac{1}{2})^{k+.5}$	Typical relatives
0	1/2 = .5000	.3536 − .7071	Monozygous co-twin
1	1/4 = .2500	.1768 − .3536	Sibs, parent-child
2	1/8 = .1250	.0884 − .1768	Uncle-niece, half-sibs
3	1/16 = .0625	.0442 − .0884	First cousins
4	1/32 = .0312	.0221 − .0442	First cousins once removed
5	1/64 = .0156	.0110 − .0221	Second cousins
6	1/128 = .0078	.0055 − .0110	Second cousins once removed
7	1/256 = .0039	.0028 − .0055	Third cousins
8	1/512 = .0020	.0014 − .0028	Third cousins once removed
9	1/1024 = .0010	.0007 − .0014	Fourth cousins

For example,

$$P(aa\,|\,aa) = C_2 + C_1 q + C_0 q^2 \qquad (7.1.3)$$
$$P(Aa\,|\,Aa) = C_2 + C_1/2 + 2C_0 q(1-q)$$

where q is the gene frequency of *a*. Li and Sacks gave, in the ITO method, a compact notation.

Thus, kinship is a single parameter not dependent on the population gene frequencies which, together with those frequencies, permits a probability statement about the genotype of an individual when a relative is known. This prediction is more reliable than we would have made if the individuals were incorrectly assumed to be unrelated or if the relative's genotype was ignored.

Table 7.1.3: Coefficients of identity for regular relatives

Relationship	Kinship (ϕ)	Identity coefficients C_2	C_1	C_0	Degree of kinship (k)
Identical twins	1/2	1	0	0	0
Sibs	1/4	1/4	1/2	1/4	1
Double first cousins	1/8	1/16	6/16	9/16	2
Parent-child	1/4	0	1	0	1
Grandparent-grandchild (=uncle-niece = half sibs)	1/8	0	1/2	1/2	2
First cousins (=great grandparent-great grandchild = great uncle-niece)	1/16	0	1/4	3/4	3
First cousins once removed	1/32	0	1/8	7/8	4
Second cousins	1/64	0	1/16	15/16	5
Unilineal	$(1/2)^{k+1}$	0	4ϕ	$1-4\phi$	k

7.2 Inbreeding

In the last section, an individual was termed *inbred* if his parents were related. The reader may have found this paradoxical, since all members of a species are "related." We must make our ideas about inbreeding more precise, and in the process we will find how the Hardy-Weinberg law can be modified to give a better description of genotype frequencies. We seek a single parameter, called the (coefficient of) *inbreeding* F, not dependent on the gene frequencies (although its range is so dependent), which will predict genotype and mating type frequencies in populations not necessarily panmictic. Whereas kinship is a probability, inbreeding is a correlation that becomes a probability only in important special cases. If genes A, a take values 0, 1, then inbreeding F is the correlation between uniting gametes.

Consider first a subpopulation all of whose members have the same inbreeding F. Then by definition of a correlation, the expected genotype frequencies are

$$P(AA) = (1 - q)^2(1 - F) + (1 - q)F \tag{7.2.1}$$

$$P(Aa) = 2q(1 - q)(1 - F)$$

$$P(aa) = q^2(1 - F) + qF$$

These become the Hardy-Weinberg frequencies if $F = 0$. The only mathematical conditions on F are imposed by

$$P(Aa) \geq 0, \quad \text{hence } F \leq 1 \tag{7.2.2}$$

$$P(aa), \quad P(AA) \geq 0, \quad \text{and so } F > \frac{-q}{1 - q}, \quad \frac{-(1 - q)}{q}$$

The most inclusive range is $-1 \leq F \leq 1$. Since F may be negative, probabilistic interpretations are special cases. We shall consider several models, superficially different, which lead to equation 7.2.1, which is therefore surprisingly useful. The concept of F has enormously simplified population genetics, and its validity under a great variety of conditions is one of the great discoveries of mathematical biology. According to it, the problem of genotype and mating type frequencies in populations is largely reduced to determining gene frequencies and inbreeding.

Usually F has a distribution among subpopulations, taking the value F_k with probability p_k. Then

$$\alpha = \sum p_k F_k \tag{7.2.3}$$

is the mean inbreeding. We may use α and F interchangeably in applications like equation 7.2.1, which are linear in F.

If gene frequencies are constant over the population, the only departure from panmixia may be due to mating of relatives. This is Wright's *genealogical* model for F, which must be positive and is in fact, the probability that a gene drawn at random from one parent be identical by descent with a gene drawn from the other parent. Thus, the inbreeding of an individual is simply the kinship of the parents.

The genealogical model is the only one for which the expected value of F is the same for different alleles. More generally, we seek the mean of F over as many alleles and loci as can be observed. In the *partitioned* model of Wahlund, a population is divided by geographic or social barriers into subpopulations within which inbreeding is negligible. They may be arbitrarily delineated or exist only abstractly, and there may be any degree of migration among subpopulations, the barriers between which may change from one generation to the next. Let q_{ik} be the frequency of allele A_i in the kth subpopulation, which makes up a fraction p_k of the total $(\sum p_k = 1)$. The mean gene frequency of A_i is $q_i = \sum_k p_k q_{ik}$ with variance $\sigma_i^2 = \sum p_k(q_{ik} - q_i)^2$. The genotype frequencies take the form of equation 7.2.1 with

$$F_i = \frac{\sigma_i^2}{q_i(1 - q_i)} \tag{7.2.4}$$

Note that $0 \le F_i < 1$, which is consistent with a probabilistic interpretation, but F_i need not be constant for all alleles. However, in genetic epidemiology we are usually concerned with small values of F_i, such as occur within an ethnic group. Since attempts to detect selection in such populations have been mostly unsuccessful, we have no reason to suppose that variation in estimates of F_i is systematic in rather homogeneous regions.

The *heirarchical* model of Wright assumes that populations may be classified into a branching structure (fig. 7.2.1). Let F_{IS} be the inbreeding of an individual relative to its locality S, F_{SR} be the inbreeding of the locality relative to a region R, and F_{RT} be the inbreeding of the region relative to a larger population T. By path analysis and later by analysis of variance, Wright showed that the inbreeding F_{IT} of an individual relative to T is given by

$$1 - F_{IT} = (1 - F_{IS})(1 - F_{SR})(1 - F_{RT})$$

or $\qquad F_{IT} = F_{IS} + (1 - F_{IS})F_{ST}$ (7.2.5)

where $\qquad F_{ST} = F_{SR} + (1 - F_{SR})F_{RT}$

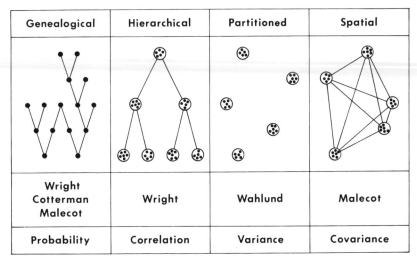

Genealogical	Hierarchical	Partitioned	Spatial
Wright Cotterman Malecot	Wright	Wahlund	Malecot
Probability	Correlation	Variance	Covariance

● Individual   Subpopulation

Fig. 7.2.1: Models of population structure with their authors and prime methods of analysis.

On this model, F_{IS} is a correlation and may be negative, whereas F_{SR}, F_{RT} are standardized variance components like equation 7.2.4 and must be positive. F_{SR} may be interpreted as the probability that two random genes in S be identical by descent from a gene in R. If the F are probabilities, equation 7.2.5 follows from the combination of independent events.

The last model we shall consider is the *spatial model* that distinguishes each of n populations, giving a square, symmetric, positive definite matrix Φ whose ij^{th} element is the kinship ϕ_{ij} between populations i and j, defined as the probability that a random gene from i be identical by descent with a random allele in j. Clearly $E(\phi_{ij}) \geq 0$. By comparison with the hierachical model, ϕ_{ij} is of the type represented as F_{ST} in equation 7.2.5 for differentiation of localities within neighboring regions.

Contrary to what the reader might suppose at this point, the definition of inbreeding is not arbitrary but is determined simultaneously with the gene frequencies. Thus, if the gene frequency q_S is estimated for a particular locality, we are only concerned with F_{IS} and estimate the frequency of the *aa* homozygote as

$$E\{P(aa)\} = q_S^2 + q_S(1 - q_S)F_{IS}$$

If the gene frequency q_R is determined for a region, then we estimate

$$E\{P(aa)\} = q_R^2 + q_R(1 - q_R)F_{IR}$$

where
$$F_{IR} = F_{IS} + (1 - F_{IS})F_{SR}$$

Finally, if the gene frequency q_T is determined for a group of regions, we take

$$E\{P(aa)\} = q_T^2 + q_T(1 - q_T)F_{IT}.$$

Genotype frequencies will deviate from expectation if there are errors in the estimate of q and F or the actual population differs from our assumptions, but each of these estimates is appropriate under the given definition of q. The best estimate of genotype frequencies is from the smallest population that gives a reliable estimate of q. This will usually have a small value of F, since $F_{IS} < F_{IR} < F_{IT}$ and $F_{SR} < F_{ST}$. We shall be concerned with the comparison of different estimates of these quantities.

However q and F are defined, the effect of inbreeding on genotype frequencies is important only when $q(1 - q)|F|$ is large relative to q^2. Since $|F|$ is small, this condition implies $q \to 0$, $1 - q \to 1$, and $P(aa) \to q[q + F]$. Therefore, inbreeding is negligible if $|F| < q$. We shall see, in the next chapter, that this conclusion applies not only to the effect of inbreeding on genotype frequencies, but also to its effect on the equilibrium gene frequency.

Biometrical genetics sometimes uses a function of the inbreeding coefficient called the (coefficient of) *relationship* R. If the kth allele is assigned a value x_k, which is additive, the correlation between genotypic values $\Sigma_{k=1}^2 x_k$ for individuals i and j with kinship ϕ_{ij} and inbreeding F_i, F_j is

$$R = \frac{2\phi_{ij}}{\sqrt{(1 + F_i)(1 + F_j)}}$$

Since F in man is generally small, R is virtually identical to $2\phi_{ij}$.

7.3 Genetic Loads

Rare recessive genes are held in the population by a balance between selection and recurrent mutation. Table 7.3.1 shows the genotype frequencies and nonaffection probability $1 - s$ in homozygotes and $1 - hs$

Table 7.3.1: The mutation load

Genotype	GG	Gg	gg
Nonaffection probability	1	$1 - hs$	$1 - s$
Frequency	$(1 - q)^2(1 - F) + (1 - q)F$	$2q(1 - q)(1 - F)$	$q^2(1 - F) + qF$

in heterozygotes. The nonaffection probability in the population is $1 - qFs - q^2(1 - F)s - 2q(1 - q)(1 - F)sh$, where qFs is the probability of affection due to homozygosity from inbreeding, $q^2(1 - F)s$ is the probability of affection due to homozygosity not from inbreeding, and $2q(1 - q)(1 - F)sh$ is the probability of affection due to heterozygosity. We must also consider the probability x of a particular environmental cause of affection. If different causes of affection act independently, then the probability of nonaffection is the product of such terms over all loci and environmental causes, or

$$S = \Pi(1 - x)[1 - qFs - q^2(1 - F)s - 2q(1 - q)(1 - F)sh \qquad (7.3.1)$$

$$\doteq e^{-(A + BF)}$$

where $A = \Sigma x + \Sigma q^2 s + 2\Sigma q(1 - q)sh$

and $B = \Sigma qs - \Sigma q^2 s - 2\Sigma q(1 - q)sh$

Then A is called the *panmictic load*, B the *inbred load*, and A + BF the *expressed load*. The *total load* Σqs lies between B and A + B lethal, sterile, or detrimental equivalents per gamete.

Similar results can be produced by the segregation load due to heterozygote advantage (table 7.3.2). In this case

$$A = \Sigma x + \Sigma q^2 s \qquad (7.3.2)$$

$$B = \Sigma qs - \Sigma q^2 s$$

which differ from the mutation load only by omission of the term in h.

The expressed load is nonlinear in F, if different loci interact or if they are linear on some other scale, such as liability. Then the affection probability is

$$S = e^{-(A + BF + CF^2 + \ldots)} \qquad (7.3.3)$$

The nonlinear component $CF^2 + \ldots$ is called the *phenodeviant load*. Under the low levels of inbreeding observable in man, the phenodeviant load is nonsignificant.

Table 7.3.2: The segregation load

Genotype	G^iG^i	G^iG^j	G^jG^j
Nonaffection probability	$1 - s_i$	1	$1 - s_j$
Frequency	$q_i^2(1 - F) + q_i F$	$2q_i q_j(1 - F)$	$q_j^2(1 - F) + q_j F$

Care is required to obtain unbiased estimates of A and B by control of environmental differences between levels of inbreeding. Neighbors and sibs who marry nonrelatives are good controls, and the data may also be stratified by liability indicator into homogeneous sets. These precautions were often not taken in the older literature. Together with small sample size and inconsistent definition of morbidity, this leads to noise in estimates of the inbred load (table 7.3.3). Expressed as mortality before reproductive maturity, the inbred load B in contemporary populations is about one lethal equivalent, and morbidity accounts for about one detrimental equivalent in addition. Under more primitive conditions, many detrimental equivalents were lethal and so the inbred load may have amounted to two lethal equivalents per gamete.

Evidence from several sources indicates that these inbreeding effects are almost entirely due to genes with a *megaphenic* effect (ie, large relative to the phenotypic standard deviation). Inbred loads for muscular dystrophy, deaf mutism, severe mental defect, and congenital malformation are restricted to recessive genes of high penetrance, since isolated

Table 7.3.3: Estimates of genetic loads from inbreeding in man.

Component	Study	A	B	B/A
Mortality	USA	0.161	1.734	10.75
before	USA	0.130	1.032	7.94
reproductive	USA	0.174	1.309	7.52
maturity	Italy	0.228	1.907	8.36
	Italy	0.213	0.918	4.34
	France	0.127	2.362	18.60
	Brazil	0.434	0.930	2.14
	Japan	0.155	0.800	5.16
Morbidity	USA	0.107	5.792	54.13
	USA	0.103	1.164	11.30
	Italy	0.024	2.880	120.00
	Italy	0.029	0.560	19.31
	Italy	0.047	0.879	18.70
	Sweden	0.082	2.020	24.63
	France	0.044	2.196	49.91
	Brazil	0.042	0.443	10.50

cases have exactly the inbreeding rate expected for a mixture of sporadic cases unaffected by inbreeding and chance-isolated cases due to major genes. This result could have been anticipated from other organisms. Inbreeding in the European bison acts on mortality through recessive lethals, not genes of small effect. In Drosophila, 70 percent of the inbred load measured at $F = 1/2$ is due to lethal chromosomes, 18 percent to severe detrimentals, and only 12 percent to mild detrimentals. Thus 88 percent of the inbred load is megaphenic.

Inbred loads due to lethals and severe detrimentals are the same in Drosophila, whether calculated from $F = 1/2$ or $F = 1$. However, the inbred load attributed to mild detrimentals increases with F, indicating that some impairment is due to synergism in multiple homozygotes. Abnormal phenotypes due to multiple homozygosity have been called *phenodeviants*. They can be unequivocally recognized only at high levels of inbreeding, and their evolutionary significance is obscure. Presumably the constituent gene frequencies are determined by first-order effects, but whether these are neutral, deleterious, or advantageous has not been amenable to study. A corollary of synergism for mild detrimentals is that their contribution to the inbred load would be even smaller if they were estimated at lower levels of inbreeding, as in man.

If synergism of multiple homozygotes were an important cause of morbidity in natural populations, then inbred loads would be largely due to mild detrimentals, which is not the case. Synthetic lethals have been reported occasionally, but the few that have been analyzed in detail involve deleterious genes, the main effects of which must be more important than their rare interactions. Since synergism is virtually restricted to a fraction of mild detrimentals, it accounts for at most a small percentage of the inbred load measured at small values of F.

Several reasons have been suggested for the small impact of synergism on the inbred load, and the unproductiveness of research on interactive systems. Some of these are the same as were used in section 5.6 against polygenic dominance and epistasis. Additional reasons are: (a) the heritability of important components of fitness is usually low, and so truncation models of the kind that have been studied extensively by mathematical and Monte Carlo methods probably have little relevance to natural populations. (b) Phenodeviants that can be detected at high levels of inbreeding are unimportant at low levels, since $\partial(A + BF + CF^2 ...)/\partial F = B + 0(F)$. (c) Interactions of common genes have virtually no effect on the inbred load. For example, suppose that a multiply homozygous phenodeviant includes at least one rare gene with frequency r, the other k more common genes having frequency Q. Then the contribution of this hypersystem to the expressed load will be

$$srQ^k[r + (1 - r)F]\{Q + (1 - Q)F\}^k = a + bF + 0(F^2)$$

where $a = sr^2Q^{2k}$ and $b \doteq srQ^{2k-1}\{(1 - r)Q + rk(1 - Q)\}$. This gives essentially the same panmictic and inbred loads as a single rare gene with frequency r and selection coefficient sQ^{2k}. (d) Phenodeviants dependent on two or more rare genes are negligible in probability, provided there is some selection against the genes in more frequent combinations. (e) Since detrimental genes approach additivity as their homozygous effects on fitness decrease, the number of generations a deleterious gene persists is similar for a mild detrimental and a recessive lethal. Therefore, in Drosophila the proportion of the inbred load due to detrimentals is about the same at equilibrium as during recovery from an enormous cumulative dose of ionizing radiation. Additive effects in heterozygotes reduce detrimental gene frequencies to about the same levels as recessive lethals, where interactions of multiple homozygotes are too rare to dominate main effects.

Any of these explanations seems sufficient to account for the negligible role of synergism on the inbred load, and the unimpressive role of synergism in nearly panmictic populations. We are, therefore, justified in estimating the frequencies of major recessive genes from inbreeding effects.

Suppose we can isolate the contributions of completely penetrant, recessive genes to give

$$A' = \Sigma q^2 \tag{7.3.4}$$

$$A' + B' = \Sigma q$$

This may use an estimate of B from inbreeding and the fact that q estimated by segregation analysis in a population with mean inbreeding α is actually $\sqrt{A' + B\alpha}$. Letting Q be the mean gene frequency and Δ^2 the variance of q, we may write

$$A' = n(Q^2 + \Delta^2) \tag{7.3.5}$$

$$A' + B' = nQ$$

Then

$$Q \leq A'/(A' + B') \pm \sigma_B/n \tag{7.3.6}$$

$$n \geq (A' + B')^2/A' \pm 2\sigma_B/Q$$

Table 7.3.4 shows that at least several recessive genes cause deaf-mutism and a large number cause mental defect.

Table 7.3.4: Estimates of mutation rate from genotypic incidence

Disease	Panmictic load $(A = \Sigma q^2)$	Inbred load $(B = \Sigma q - \Sigma q^2)$	Selection coefficient (m)	Mutation rate per gamete $(u \times 10^6)$	Number of loci $(n \geq (A + B)^2/A)$	Mutation rate per locus $(u \times 10^6)$
Deaf mutism	.000180	.080	.68	545	36	15
Limb girdle muscular dystrophy	.000033	.008	.75	60	2	31
Severe mental defect	.000324	.192	1.00	1923	114	17
All mental defect	.001964	.830	.75	6240	352	18
Retinal degeneration	.00021	.047	.15	70	10	7
Retinitis pigmentosa	.00022	.022	.15	34	2	15

A check on the estimate for mental retardation is provided by inbreeding effects on IQ. Since genotype frequencies for a single locus are linear in F, the effect of inbreeding on the mean of a quantitative trait is expected to be linear if there is no interaction between loci. The mean regression is $b = -44.0 \pm 12.3$. Let us suppose that this is entirely due to rare, recessive genes with displacement t and frequency Q per locus. Taking $t = 3.5$ from segregation analysis and $Q = .0025$ from inbreeding analysis, we estimate

$$n \geq -b/tQ \tag{7.3.7}$$

which in this example is 335. We could hardly ask for better agreement with the estimate of 352 in table 7.3.4, which supports the assumption that the decline of IQ with inbreeding is entirely due to rare, recessive genes.

Estimation of the inbred load B is not limited to random samples in which the equation $S = e^{-(A+BF)}$ may be fitted. Selected samples of affected individuals are also suitable. Let c_i be the frequency of inbreeding F_i in the general population, grouping F_i by degree as in table 7.1.2. The mean is $\alpha = \Sigma c_i F_i$ with variance $\sigma^2 = \Sigma c_i F_i^2 - \alpha^2$. For rare traits and small values of F_i we may write the affection frequency as $I \equiv 1 - S = 1 - e^{-(A+BF_i)} \doteq A + BF_i$. Then among probands the mean inbreeding is

$$F = \frac{\Sigma c_i F_i [A + BF_i]}{\Sigma c_i [A + BF]} \tag{7.3.8}$$
$$= \frac{A\alpha + B(\sigma^2 + \alpha^2)}{I}$$

Solving these simultaneous equations for I and F in terms of A and B we obtain

$$B = I(F - \alpha)/\sigma^2$$
$$\tag{7.3.9}$$
$$A = I - B\alpha$$

Fully efficient estimates may be obtained from the likelihood of m_i probands with F_i,

$$L = \prod_i c_i^{m_i} \left\{ \frac{A + BF_i}{A + B\alpha} \right\}^{m_i} \tag{7.3.10}$$

7.4 Evolution of Kinship

We have seen that the phenomenon of inbreeding is complex, since an individual belongs simultaneously to a small local population and to successively larger regions, for each of which there is a vector of gene frequencies and corresponding inbreeding.

Faced with this complexity the geneticist has no choice but to over-simplify. An important special case of all four models of figure 7.2.1 is the genetic island, a population of effective size N and linear pressure m which are constant from generation to generation. In such a population kinship increases up to a certain point, after which it stabilizes.

The *effective population* of a genetic island is defined by its gene frequency variance σ_q^2 each generation,

$$N \equiv \frac{q(1-q)}{2\sigma_q^2} \tag{7.4.1}$$

Many special cases have been studied mathematically, leading to the generalization that effective size is approximately one-third the harmonic mean of census size, which includes children and elderly adults.

The systematic pressure on a genetic island is defined by the expected change in gene frequency per generation,

$$m = E\{(q_t - q_{t-1})/(Q - q_{t-1})\} \tag{7.4.2}$$

where $E(q_t) = (1-m)q_{t-1} + mQ$. This definition is satisfied if a gamete in generation t comes from the previous generation with probability $1-m$ and from an unchanging pool of migrants with probability m, where q is the gene frequency in the island and Q is the frequency in the migrant pool.

We are now ready to derive the evolution of kinship in a genetic island. On the genealogical model kinship in generation t is the probability that two random genes be identical by descent, which is

$$\phi^{(t)} = (1-m)^2 \left[\frac{1}{2N} + \left(1 - \frac{1}{2N}\right)\phi^{(t-1)} \right], \qquad t = 1, 2, \ldots$$

Rearranging,

$$\phi^{(t)} - \phi^{(t-1)} = (1-m)^{2t}(1 - 1/2N)^{t-1}/2N$$

$$\doteq \frac{1}{2N} e^{-(2m + 1/2N)t}$$

By integration,

$$\phi^{(t)} \doteq \frac{1}{4Nm + 1} [1 - e^{-(2m + 1/2N)t}]$$

(7.4.4)

The equilibrium value of kinship is

$$\phi^{(\infty)} \doteq \frac{1}{4Nm + 1}$$

(7.4.5)

and the number of generations to go halfway to equilibrium is

$$t \doteq \frac{\ell n2}{2m + 1/2N}$$

(7.4.6)

These results give a remarkably good description of random kinship not only within a genetic island, but even in complex structures under constant forces of drift and systematic pressure. The parameters are the evolutionary size N_e (which differs from effective size N by partial inclusion of populations which contribute immigrants) and the effective systematic pressure m_e (which differs from the systematic pressure m by partial inclusion of immigrants from within the structure).

To validate this result we must consider a region of n populations that generate the $n \times n$ matrix whose typical element p_{ij} is the probability that an individual having a child in population j was born in population i. Then $\Sigma_{i=1}^n p_{ij} = 1$. Associated with this matrix are vectors of effective sizes N_i and systematic pressures m_i from immigrants born outside the region and assumed to come from an unchanged gene pool. Then random kinship between populations i and j in generation t is

$$\phi_{ij}^{(t)} = (1 - m_i)(1 - m_j)\left\{ \sum_{k=1}^{n} \sum_{h=1}^{n} p_{ki} p_{hj} \phi_{hk}^{(t-1)} \right.$$

(7.4.7)

$$\left. + \sum_{k=1}^{n} p_{ki} p_{kj}(1 - \phi_{kk}^{(t-1)})/2N_k \right\}$$

The $n \times n$ kinship matrix $\phi^{(t)}$ is square, symmetrical, and reaches an equilibrium between drift and systematic pressure. Successive terms $\phi_{ij}^{(1)}$ . . ., $\phi_{ij}^{(t)}$ follow equation 7.4.4 to a close approximation, with parameters N_e, m_e that can be estimated.

The transition probabilities p_{ij} are defined on parent-child pairs, which have a simple relation to pairs of mates if at least one mate was born in the population of residence. On this assumption, let m_{ij} be the probability that a male born in population i have a wife from population j. Then

$$m_{ij} = \begin{cases} 2p_{ij} & i \neq j \\ 2p_{ii} - 1 & i = j \end{cases} \tag{7.4.8}$$

The mean inbreeding coefficient in population i is

$$\alpha_i^{(t)} = \sum_{j=1}^n m_{ij} \phi_{ij}^{(t)} = 2 \sum_{j=1}^n p_{ji} \phi_{ij}^{(t)} - \phi_{ii}^{(t)} \tag{7.4.9}$$

we may use this result by noting that the increase of inbreeding from generation $t - 1$ to t corresponds to $F_k = (1/2)^{2t}$, and so the predicted frequency of F_k is

$$p_k = 2^{2t}(\alpha^{(t)} - \alpha^{(t-1)}) \tag{7.4.10}$$

where $k = 2t - 1$ is degree of kinship.

Suppose that *close inbreeding* up to and including generation t is separately distinguished (usually $t = 2$ or 3), and for a given mating takes the value $F_c = 0, 1/2, \ldots (1/2)^{2t}$. Then for a child born in population i, the total inbreeding is

$$F_{IT} = F_c + (1 - F_c)F_r \tag{7.4.11}$$

where *remote inbreeding* is

$$F_r \equiv (\alpha_i^{(\infty)} - \alpha_i^{(t)})/(1 - \alpha_i^{(t)}), \tag{7.4.12}$$

the inbreeding due to common ancestors more than t generations back. Substituting $F_c = \alpha_i^{(t)}$ gives $F_{IT} = \alpha_i^{(\infty)}$ as expected.

The corresponding equation if one parent is known to come from population i and the other parent from j is

$$\phi_T = \phi_c + (1 - \phi_c)\phi_r \tag{7.4.13}$$

where ϕ_c, ϕ_r are close and remote kinship, respectively.

The description of population structure given by migration is detailed but subject to two sources of error. First, it does not allow for avoidance of close inbreeding or preferential consanguineous marriage within populations. Secondly, the data are usually imperfect, with uncertainty in the estimates, temporal changes, and missing observations.

Equality in numbers of reciprocal migrants must be imposed if population sizes are to be stable at their assumed frequencies. Missing observations may be estimated, but we need to verify predictions of kinship and inbreeding from migration (table 7.4.1).

Table 7.4.1 : Means of inbreeding parameters

Population	Number of samples	$10^5 \, \alpha_\infty$	α_2/α_∞	N_e	m_e	Harmonic mean N_e	m_e
Middle East	20	1524	.76	175	.441	34	.282
Isolate controls	31	980	.44	864	.177	139	.102
National populations	18	176	.49	3936	.212	862	.125

Genealogies provide independent information from which close inbreeding can be estimated. Starting with founders, mean inbreeding $\alpha(t)$ in successive generations gives total inbreeding by equation 7.4.4. In the most recent generation degrees of kinship may be tabulated. Let n_k be the number of instances of degree k, each composed of two chains of length $k + 1$. Thus first cousins (who are of degree 3) correspond to two chains of length $k = 4$. If N is the number of pairs on which kinship is based, then the cumulative kinship in generation t is

$$\phi(t) = \sum_{k=1}^{2t-1} n_k (1/2)^{k+1}/N \qquad (7.4.14)$$

Large values of k that are not reliably ascertained should be omitted when fitting this estimate of $\phi(t)$ by equation 7.4.4. Applied to pairs of mates it gives inbreeding. Applied to random pairs of individuals, it gives kinship.

7.5 Bioassay of Kinship

We have seen how genealogy and migration give predictions of kinship which require inductive confirmation. This is called *bioassay of kinship*, which subsumes all attempts to estimate kinship from discrete phenotypes, quantitative traits, clans, or surnames.

The simplest approach is to estimate gene frequencies and inbreeding simultaneously from phenotype frequencies (equation 7.2.1). Any classification error or misspecification of genetic systems will distort this estimate of inbreeding, which is slightly biased in small samples. Yasuda showed that samples of mating pairs give about six times as much information as for parental phenotypes considered individually. His expression for mating type frequencies ignores polynomials in F, and is therefore appropriate if inbreeding is less than the smallest gene frequency (table 7.5.1).

This method does not extend to random pairs of individuals between populations, where inbreeding and random kinship are confounded. We therefore turn to pairs of gene frequency vectors, treating a sample of size n_i as drawn from an infinite genetic pool. Let $\hat{q}_{ir}$ be the maximum likelihood estimate of the frequency of the rth allele in the ith population, where Q_r is the regional mean. Then information theory provides estimates for the kth system with m alleles.

$$\hat{\phi}_{ij} = \begin{cases} \dfrac{J_{ii}/M - 1/2n_i}{1 - 1/2n_i} & i = j \\ \\ J_{ij} & i \neq j \end{cases} \tag{7.5.1}$$

where

$$J_{ij} = \sum_{r=1}^{m-1} \sum_{s=1}^{m-1} K_{rs}(\hat{q}_{ir} - Q_r)(\hat{q}_{is} - Q_s)$$

is called the *codivergence* between populations i and j. The weight K_{rs} is an element in the information matrix per individual about gene frequencies Q_r, Q_s, and

$$M = \sum_{r=1}^{m-1} \sum_{s=1}^{m-1} K_{rs} \delta_{rs} \tag{7.5.2}$$

$$\delta_{rs} = \begin{cases} Q_r(1 - Q_r) & r = s \\ -Q_r Q_s & r \neq s \end{cases}$$

With small samples from a rather homogeneous region the maxi-

Table 7.5.1: Mating type frequencies with two alleles, A and a, in frequencies p and $q = 1 - p$

Mating type	Probability
AA × AA	$p^4 + 6p^3qF$
AA × Aa	$4p^3q + 12p^2q(1 - 2p)F$
AA × aa	$2p^2q^2 + 2pq(1 - 6pq)F$
Aa × Aa	$4p^2q^2 + 4pq(1 - 6pq)F$
Aa × aa	$4pq^3 + 12pq^2(1 - 2q)F$
aa × aa	$q^4 + 6pq^3F$

$$\left.\phantom{\begin{array}{c} p^4 \\ 4p^3q \\ 2p^2q^2 \\ 4p^2q^2 \\ 4pq^3 \\ q^4 \end{array}}\right\} + 0(F^2)$$

mum likelihood estimates $\hat{q}$ are unreliable, and maximum likelihood scores evaluated at the regional gene frequencies give a better estimate of codivergence.

A vector of m quantitative traits follows equation 7.5.1 with the population means substituted for gene frequencies and

$$M = \sum_{r=1}^{m} \sum_{s=1}^{m} V_{rs}^{-1} G_{rs} \qquad (7.5.3)$$

where V is the pooled covariance matrix within groups and G is the matrix of genetic covariances, assuming that environment and measurement errors are random among populations. Since that assumption is doubtful and G is usually unknown, quantitative traits give imprecise estimates of kinship.

Surnames and clans are obviously not genetic, but they give some insight into population structure. If q_{ki} is the frequency of the kth surname in population i, then

$$I_{ij} = \Sigma q_{ki} q_{kj} \qquad (7.5.4)$$

is called *random isonymy* between populations i and j. If n_{ki} is the number of individuals with the kth surname in a random sample from the ith population where $N_i = \Sigma_k n_{ki}$, then an unbiased estimate is

$$I_{ij} = \begin{cases} \dfrac{\Sigma_k n_{ki}(n_{ki} - 1)}{N_i(N_i - 1)} & i = j \\ \dfrac{\Sigma_k n_{ki} n_{kj}}{N_i N_j} & i \neq j \end{cases} \qquad (7.5.5)$$

Since a proportion 4ϕ of common relatives with kinship ϕ are isonymous by descent, $I_{ij}/4$ is an estimate of ϕ_{ij} under three stringent assumptions: (a) all names are *monophyletic* (ie, due to descent from a common ancestor), (b) mutation and selection are negligible relative to migration, and (c) names are subject to the same forces of drift and migration as are genes.

These assumptions are violated by polyphyletic surnames, given contagiously to members of a community who may be distantly related, if at all. Names tend to change in the direction of the prevailing language, as Müller to Miller. Two clans may have fictive homology based on a desire to recognize clansmen and avoid or practice endogamy, rather than any established tradition of common descent. Transmission of names may be irregular, altered by adoption or conquest, or there may be frequent orthographic variations. Isonymy does not require painstaking study of phenotypes and permits historical studies of drift, but the hypothesis that genes and surnames are equivalent must be tested, not accepted on faith.

7.6 Isolation by Distance

Malécot has shown that random kinship as a probability approximates

$$\phi_{ij} \doteq \phi(d) = ae^{-bd} \qquad (7.6.1)$$

where
$$\phi_{ii} \doteq a = 1/(4N_e m_e + 1)$$
$$b = \sqrt{2m_e}/\sigma'$$

Here d is the distance between populations i and j, N_e is the evolutionary size of a local population, m_e is its effective systematic pressure, and σ' is the standard deviation in the plane of parent-offspring distances (excluding long-range migration). The derivation assumes equilibrium between drift and migration and the same values of N_e, m_e, and σ' for all populations. Given predictions of kinship from migration, the topological parameters a and b may be estimated for each population or overall. In practice, these parameters vary among populations, with a large for small, isolated populations and b large for sedentary ones.

When kinship is bioassayed from phenotypes or quantitative traits, it is expressed relative to random pairs of genes within the region, or

$$\phi(d) = (1 - L)ae^{-bd} + L \qquad (7.6.2)$$

where $L \leq 0$ is the kinship at large distance within the region. The mean of ae^{-bd} is random kinship in the region ($\bar{\phi}$). The mean of conditional kinship $\phi(d)$ is zero, since random pairs of genes correspond to panmixia at the regional gene frequency. Rearranging,

$$\bar{\phi} = -L/(1 - L) \qquad (7.6.3)$$

Isolation by distance provides a first-order correction to estimates of kinship from isonymy. Recall that the frequency of isonymy by descent is $i = 4\phi$ for common relatives (eg, $i = 1$, $\phi = 1/4$ for sibs and $i = 1/4$, $\phi = 1/16$ for first cousins). Then we may write the frequency of isonymy at distance d as

$$I(d) = (1 - L)i + L \qquad (7.6.4)$$

$$= 4(1 - L)ae^{-bd} + L$$

where L is the frequency of random isonymy at large distance and is therefore positive. Polyphyletic names like Smith, Jones, and Woods, which are taken randomly in different places, contribute exclusively to L, and therefore

$$\phi_{ij} = \frac{I_{ij} - L}{4(1 - L)} \qquad (7.6.5)$$

is a better (and smaller) estimate of kinship than the crude estimate $I_{ij}/4$ from equation 7.5.4, which includes random polyphyletic names. This estimate is still contaminated by locally polyphyletic names, assumed by several people in the same community, and so estimates of kinship by equation 7.6.5 tend to be biased upwards. However, the history of surnames is short relative to genes, and this biases such estimates downwards. We should be cautious in accepting estimates of kinship from isonymy (table 7.6.1).

Bioassay of F_{IS} from marital isonymy within a local population is immune to this criticism but depends on two strong assumptions: first, surnames before marriage are correctly ascertained without occasional substitution of married names, and second the factor of 4, which applies to random pairs, is conserved in matings. Any tendency to prefer certain sex-labeled types of consanguineous marriage (such as children of two brothers) will alter this factor which, however, has been found to ap-

Table 7.6.1 : Isonymy in different populations

Population	$\dfrac{\Sigma q^2}{4}$	Marital isonymy (I_m)	$F = \dfrac{I - \Sigma q^2}{4(1 - \Sigma q^2)}$	Pedigree estimate of F
Caucasians	.00032	—	—	—
Japanese	.00051	—	—	—
Hawaiians	.00029	—	—	—
Filipinos	.00028	—	—	—
Chinese + Koreans	.00660	—	—	—
All races, Hawaii	.00027	—	—	—
England, 19th century	.00026	.0125	.0029	.0024
Colonial New York	.00020	.0207	.0050	.0055
Ohio, 19th century	—	.0084	.0019	.0018
Loomis genealogy	—	.0076	.0017	.0030
Burke's landed gentry	—	.0151	.0037	.0035
Japan, consanguineous marriages	—	.1560	.0387	.0466
Austria, consanguineous marriages	—	.1590	.0395	.0507
Hutterites	.04459	.1951	.0052	—

proximate 4 in several populations. On these assumptions, the frequency of marital isonymy is

$$I_m = 4F_{IS} + (1 - 4F_{IS})\Sigma q^2 \qquad (7.6.6)$$

and so,

$$F_{IS} = \frac{I_m - \Sigma q^2}{4(1 - \Sigma q^2)} \qquad (7.6.7)$$

where Σq^2 is the frequency of random isonymy in the isolate.

However the topological parameters a and b may be estimated, interpretation in terms of migration encounters two difficulties. First, the distinction between long-range and short-range migration is arbitrary. It has been found that if σ^2 is taken as $E(d^2)$ with all migrants included, then consistency with other evidence is obtained if $d > 4\sigma$ is defined as "long-range". Then σ' is the standard deviation with these migrants excluded. Much of the mathematical theory is based on stepping-stone models in which all short-range migrants come from a small number of populations in the region. If $d < \sigma'/10$ defines a locality, the estimates are consistent with Malécot's approximation,

$$m_e \doteq \sqrt{m(m + 2k)} \qquad (7.6.8)$$

where m is the frequency of long-range migrants and k is the frequency of short-range migrants $(d > \sigma'/10)$ when long-range migrants are excluded.

Whereas the kinship of many populations has been bioassayed, values of the demographic parameters N_e, m_e, and σ' are known for a smaller number, which give the following approximations:

$$\ell n\ \sigma' \doteq .060 - .728\ \ell n\ b \qquad (7.6.9)$$

$$\ell n\ N_e \doteq 2.666 - .667\ \ell n\ a$$

$$\ell n\ m_e \doteq -4.005 - .304\ \ell n\ a$$

They are less reliable than estimates from migration and genealogy, but for populations with contemporary dispersal and admixture or poorly documented histories, they provide tentative inference of the forces that produced the structure (table 7.6.2).

A population may be considered an *isolate* if it is appreciably differentiated from others of the same ethnic group by genetic drift, which

Table 7.6.2: Hierarchic kinship in Northeastern Brazil

Subregion	Number in Northeast n	Effective size N_e	Mean squared d σ^2	Corresponding $\dfrac{d}{\sqrt{\sigma^2}}$	Mean co-efficient of inbreeding within region, α	Mean co-efficient of kinship among regions, $\varphi(d)$	Total $\varphi + (1 - \varphi)\alpha$
Nordeste	1	7,000,000	4,400,000	2,100	.0080	0	.0080
Estado	7	1,000,000	630,000	790	.0074	.0006	.0080
Neighborhood	27	260,000	160,000	400	.0066	.0014	.0080
Município	1,083	6,500	4,100	64	.0051	.0029	.0080
Distrito	2,725	2,600	1,600	40	.0049	.0031	.0080

implies $a > .005$. Members of an isolate tend to mate endogamously, the effective immigration rate is small, and internal geographical barriers to migration are not important. With continuous populations this leads to a large value of b, but if populations exchange migrants over water or uninhabitated territory, b may be small. Isolates are aggregated into regional and national populations, which are still sufficiently differentiated internally, or from their neighbors, to provide special opportunities for genetic epidemiology.

To understand a particular isolate, its contemporary gene frequencies must be interpreted in terms of past events. R. A. Fisher remarked: "While genetic knowledge is essential for the clarity it introduces into the subject, the causes of the evolutionary changes in progress can only be resolved by an appeal to sociological and even historical facts," which are known only for man and the organisms of which he is a vector. Gajdusek goes even farther: "When exact historical data of the precise matings which gave rise to any pattern of gene distribution in small groups are known, no mathematical calculations which might predict the extent of possible 'drift' are any longer entertained. The historical facts 'explain' fully the pattern observed."

Favorable cases are not far from this prescription. Pedigrees of the Old Order Amish have been reconstructed from genealogical records back to prerevolutionary immigrants. Genealogies for Pingelap and Mokil atolls have been drawn up from oral tradition since 1775, when a typhoon reduced the population to a small group of founders. The Ramah Navaho, Tristan Da Cunha, and the Cocos Islands represent other populations where fairly complete genealogies can be reconstructed for more than one, but less than three, centuries. Exceptional pedigree depth is associated with uncertain reliability of the early generations.

Longer time periods impose two restrictions: ascertainment of the genealogy deteriorates to the point where, at most, single lines of descent for important persons can be traced, and the historical record omits genetically significant events. How did African genes enter the Habbanites and the Jebeliya, whose history as distinct groups goes back more than 1000 years? Is the focus of Tay-Sachs disease and other rare genes in East European Ashkenazi Jews due to genetic drift, as Fraikor has argued from the history of drastic population contraction and expansion for hundreds of years? Rao and Morton calculated that at least a few among the many recessive genes in man would experience large deviations under such conditions, but this falls far short of the "exact historical data of the precise matings" Gajdusek advocates. Isolation by distance is appropriate when such data are lacking (table 7.6.3).

Table 7.6.3 : Isolation by distance (km)

Population	Source	Observed		Predicted			
		a	b	σ'	N_e	m_e	$N_e m_e$
Ethnic groups	Phenotypes	.1582	.0008	191	49	.032	2
Pingelap/Mokil atolls	Metrics	.0884	.0069	40	73	.038	3
Bougainville villages	Phenotypes	.0765	.1050	5	80	.040	3
Australian aboriginals	Phenotypes	.0641	.0013	134	90	.042	4
Pingelap/Mokil atolls	Migration	.0606	.0120	27	93	.043	4
Bougainville villages	Migration	.0588	.0954	6	95	.043	4
Micronesian atolls	Metrics	.0569	.0016	115	97	.044	4
Pingelap/Mokil atolls	Phenotypes	.0565	.0069	40	98	.044	4
Pingelap/Mokil atolls	Clans	.0540	.0111	28	101	.044	4
Micronesian atolls	Phenotypes	.0463	.0023	88	112	.046	5
New Guinea villages	Phenotypes	.0444	.0519	9	115	.047	5
Marshall atolls	Migration	.0431	.0005	269	117	.047	6
Makiritare villages	Phenotypes	.0423	.0440	10	119	.048	6
South American Indian tribes	Phenotypes	.0379	.0032	70	128	.049	6
Papago localities	Phenotypes	.0182	.0661	8	208	.062	13
Aland Is.	Migration	.0156	.1539	4	231	.065	15
Orkney Is.	Phenotypes	.0130	.0079	36	261	.068	18
Barra Island	Phenotypes	.0104	.0058	45	302	.073	22
Barra Island	Migration	.0076	.0284	14	373	.080	30
Alpine Switzerland communes	ABO groups	.0069	.0643	8	397	.083	33
Lewis Island	Phenotypes	.0061	.0057	46	432	.086	37
Northeastern Brazil localities	Phenotypes	.0050	.0064	42	493	.091	45
Oxfordshire villages	Migration	.0048	.8620	1	506	.092	47
Barra Island	Isonymy	.0048	.0112	28	506	.092	47
Sweden localities	Phenotypes	.0033	.0053	48	650	.104	67
Switzerland communes	ABO groups	.0025	.0185	19	782	.113	88
Belgium communes	ABO groups	.0009	.0247	16	1546	.154	238
Japan prefectures	Phenotypes	.0006	.0064	42	2027	.174	352

7.7 Distribution of Rare Genes

When the genealogical record is adequate, we may identify those founder individuals who carried alleles that drifted to unusually high frequency among descendants. Ideally only one couple is a common ancestor to all gene carriers, but many noncarriers are not descendants of that couple. Thus Samuel King and his wife are ancestral to 80 parents of Ellis-van Creveld patients and an unspecified proportion of noncarriers among Old Order Amish.

This principle may be given a statistical basis by reduction to a 2×2 contingency table of carrier status by ancestry. Although inter-relatedness makes a χ^2 test inexact, the extreme disproportionality leaves no doubt that all carriers in the Eastern Carolines received the achromatopsia gene from Mwahuele, the chief at the time of the great typhoon 200 years ago (table 7.7.1 and fig. 7.7.1).

The above approach uses pairs of relatives, an extremely convenient but approximate way to summarize pedigree information. It depends only on Malécot's kinship ϕ, without requiring more complex measures of joint identity by descent. This simplification is obtained by considering panmictic founders and parents of affected homozygotes, rather than the patients themselves. Information sacrificed for simplicity by dichotomizing kinship into positive or negative may be recovered in parents with kinship to the carrier founder, but they are better exploited by pedigree analysis, given preliminary evidence from the 2 × 2 contingency table. This logic has also been applied to conditions of complex or doubtful inheritance, and failure to detect a carrier founder for leprosy on Pingelap is part of the evidence that genetic susceptibility does not explain the focus of leprosy there, which is due to contacts in the phosphate mines of Nauru at the beginning of this century.

Workman has suggested that this methodology be extended to interpopulational comparisons where kinship is based not on genealogies but on phenotype bioassay or linguistic or cultural evidence. He provided a heuristic example in which discrepancy between interpopulational kinship and the prevalence of diabetes argues for cultural rather than genetic differences. Ideally such circumstantial evidence should be supported by critical intrapopulational studies, which are typified by path analysis to resolve biological and cultural inheritance and segregation analysis to discriminate major genes.

The distribution of a rare gene in space can be studied in two ways. In the first, the objective is to determine the center of dispersal and the variance of dispersion, assuming one major point of origin.

If e, f are the geographic coordinates of the center of dispersal, then the probability that an unselected individual with coordinates (X, Y) be a

Table 7.7.1: Kinship of carrier sibships to sibship 1067 for achromatopsia on Pingelap and Mokil atolls

| | Kinship to 1067 | | |
	No	Yes	Total
Not known carrier	363	727	1,090
Known carrier	0	117	117
Total	363	844	1,027

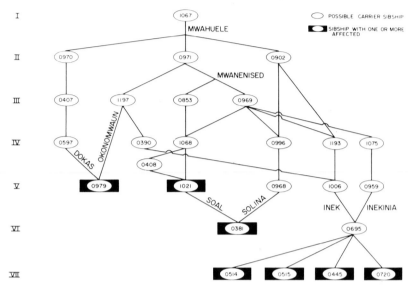

Fig. 7.7.1: Ethnohistory of achromatopsia on Pingelap atoll. All cases trace back through both parents to sibship 1067, confirming recessive inheritance.

carrier is approximately

$$P_{XY} = d + kC_{XY} \qquad (7.7.1)$$

where

$$C_{XY} = \frac{1}{2\pi V} e^{-[(X-e)^2+(Y-f)^2]/2V}$$

If an individual is selected through kinship ϕ to a proband, the probability that he be a carrier is approximately $d + 2\phi kC_{XY}$. A proportion $d/(d+k)$ of carriers come from a uniform distribution in the region, while the remainder have an isotropically normal distribution around the center of dispersal with variance $V = t(\sigma')^2$, where t is the number of generations since the beginning of dispersion and $(\sigma')^2$ is the variance of short-range migrants. This theory can be used to advantage with such diseases of restricted distribution as Tay-Sachs in Ashkenazi Jews or acheiropody in Brazil.

The second approach writes the probability that a spouse of a carrier be a carrier from equation 7.1.3 as

$$P(Aa\,|\,Aa) \doteq 2[q + \phi(d)] \qquad (7.7.2)$$

Then the distribution of marital distance is

$$r(d) = \frac{(Q + ae^{-bd})\mu(d)}{Q + \phi}, \qquad Q \doteq A/B + L \tag{7.7.3}$$

where $\mu(d)$ is the distribution of marital distances in the general population and $\phi = \Sigma\phi(d)\mu(d)$ is the mean inbreeding. The same equation holds for random pairs of parents of affected, with $\mu(d)$ as the distribution of random pairs and a larger value of Q if all affected are not homozygous for the same gene.

When this logic was applied to retinitis pigmentosa in Switzerland, the estimate of Q was significantly greater (.017) for isolated cases than for familial cases (.001), confirming evidence from segregation and consanguinity analysis that many isolated cases are nonrecessive, and may well be nongenetic. Since B is known from consanguinity analysis and L from kinship bioassay, estimation of Q gives the proportion of isolated cases that are nonrecessive and the number of loci that contribute to retinal degeneration, unlike the genetic load theory, which provides only limits.

The same logic was applied to random pairs of parents of probands, using pairs between different diseases as the control sample. Random pairs within the diseases myotonic dystrophy, hemophilia A, and hemophilia B gave estimates of Q consistent with the mean gene frequency. There is a conspicuous excess of pairs at short distances, reflecting local identity by descent. When familial retinal degeneration was divided into diagnostic groups, the estimate of Q for pairs within groups was significantly less than the total gene frequency. Thus there is no doubt that more than one locus contributes to retinal degeneration, and that the genetic basis of the diagnostic categories is in part different. Pairs of parents between diagnostic categories give $Q = .73$, indicating no appreciable identity by descent. In other words, each contributory allele tends to produce a characteristic diagnostic type, and so the clinical classification has genetic validity. Isolation by distance complements other evidence and deserves to be used more widely in genetic epidemiology.

7.8 Outcrossing

Outcrossing means that mates come from different populations. This may have two genetic effects: it will certainly decrease the frequency of homozygosity for deleterious recessive genes, and it may possibly create new genotypes, not yet subjected to the sieve of natural selection. Such disruption of coadapted genotypes is frequent in interspecific crosses but has not been demonstrated in man, where hybrids are intermediate in size, mortality, and morbidity between the parental groups.

The simplest treatment of outcrossing expresses genotype frequencies relative to a specific F_2 hybrid. Since our interest is primarily in rare homozygotes, let us follow their frequency. If the gene frequency is q in the F_2, it was $q \pm \Delta$ in the parental populations. Assuming local panmixia, table 7.8.1 shows that the difference in homozygosity between the P and F_1 generations is $2\Delta^2$, which by equation 7.2.4 equals $2q(1 - q)\alpha$, where α is the mean inbreeding in parents. If we estimate as M the difference in morbidity between parents and F_1, and as B the inbred load within populations, then

$$\alpha = -M/2B \tag{7.8.1}$$

estimates inbreeding in parental groups relative to an F_2 hybrid. The same formula with the sign changed holds for measurements, since reduction in homozygosity is expected to decrease morbidity and increase size.

When this theory was applied to 180,000 births in Hawaii, estimates of α by equation 7.8.1 were .0005 for minor hybridity and .0009 for major hybridity, where outcrossing between Caucasoids and Pacific populations was considered major and outcrossing between ethnic groups within these categories was designated minor. Neither estimate is significantly different from zero, nor are the effects of maternal hybridity and recombination significant. At the present time, human populations do not represent coadapted genetic combinations that are disrupted by outcrossing.

The small estimates of α from morbidity and size contrast with much larger estimates from polymorphisms, where bioassay of kinship has indicated that ϕ_{RT} from major racial groups is .15. Clearly polymorphisms are subject to weaker and/or more variable selection than the rare, recessive genes that contribute to outcrossing effects on morbidity.

Table 7.8.1: Homozygosity in outcrosses

Generation	P(aa)
Parents	$\frac{(q + \Delta)^2 + (q - \Delta)^2}{2} = q^2 + \Delta^2$
F_1	$(q + \Delta)(q - \Delta) = q^2 - \Delta^2$
F_2	q^2

Bioassay of kinship from polymorphisms (and a fortiori from isonymy) is relevant to rare recessive genes only over rather homogeneous regions, not the ethnic and species arrays that are of interest to evolutionary genetics.

When rare recessive diseases are considered, outcrossing effects may be striking. For example, cystic fibrosis has a gene frequency of .016 in Caucasians of Western European ancestry, but only .003 in non-Caucasians. The Caucasian incidence of 26×10^{-5} is reduced to 5×10^{-5} in outcrosses. For this selected gene the estimate of α from equation 7.2.4. is .0043. The mean value of α for unselected rare recessive genes is certainly smaller, consistent with the estimate of .0009 for major outcrosses in Hawaii. Even such a small value has a relatively large effect on the incidence of disease due to rare recessive genes, which from table 7.8.1 is about $(q - \alpha)/(q + \alpha)$ as great in outcrosses as in incrosses.

The theory of outcrosses leads to a measure of genetic distance with an interesting property. Consider an outcross between two populations i and j as a random pair from a larger set of populations, in which the gene frequency is Q. Relative to the larger population, the F_2 has kinship $\phi = (\phi_{ii} + \phi_{jj} + 2\phi_{ij})/4$. The kinship of parents relative to this F_2 is a positive correlation defined as *hybridity*,

$$\theta_{ij} \equiv \frac{\phi_{ij} - \phi}{1 - \phi} = \frac{\phi_{ii} + \phi_{jj} - 2\phi_{ij}}{4 - \phi_{ii} - \phi_{jj} - 2\phi_{ij}} \tag{7.8.2}$$

This measure of genetic distance has zero expectation if $i = j$ and increases to a logical maximum of unity when the populations are fixed for different alleles (fig. 4.8.1). Therefore hybridity expresses the proportion of gene substitutions between populations i and j. The D^2 statistic of Mahalanobis and other common measures of distance are proportional to $\phi_{ii} + \phi_{jj} - 2\phi_{ij}$, and therefore are nearly proportional to hybridity.

Substituting from equation 7.6.1, hybridity as a function of distance is

$$\theta(d) = \frac{a(1 - e^{-bd})}{2 - a(1 + e^{-bd})} \tag{7.8.3}$$

which has the remarkable property of being independent of L. Hybridity is therefore convenient for estimating the Malécot parameters a and b.

Efficient estimates of the proportions of racial admixture from phenotype frequencies are given by maximum likelihood when the ancestral gene frequencies are known with negligible error (table 7.8.2). This assumption may be acceptable for major races, but not for minor ethnic

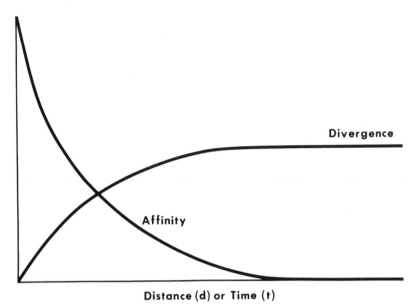

Fig. 7.8.1: Affinity and divergence. Kinship and correlation measure affinity. Hybridity and genetic distance measure divergence.

groups. Then for two ancestral populations (C, N) kinship and hybridity give estimates

$$m_i = (\theta_{CN} + \theta_{CI} - \theta_{NI})/2\theta_{CN} \qquad (7.8.4)$$

$$z_i = (\theta_{CI} - m^2\theta_{CN})(4 - \phi_{ii})$$

where m_i is the proportion of group N in the ancestry of population i and z_i is the local kinship due to drift. Unless the ancestral populations C and N were appreciably differentiated, the estimate of m_i is unreliable and z_i is approximately equal to ϕ_{ii}, indicating that local differentiation is due largely to drift and not to heterogeneity in ethnic composition.

7.9 Topology

The square, symmetrical kinship matrix Φ encodes information about relationship of n populations, which can be summarized (with some loss of detail) in 1 or 2 dimensions. Such a representation of kinship in a small number of dimensions is called *topology* of kinship. Its purpose is to reveal similarities as a guide to genetic history.

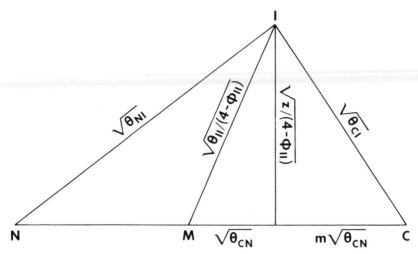

Fig. 7.8.2: Hybridity and admixture, showing a local population (I), Celtic ancestors (C), Norse ancestors (N), and the regional gene pool (M).

A common method uses a centroid adjustment,

$$\phi'_{ij} \equiv \phi_{ij} - \phi_{i.} - \phi_{.j} + \phi_{..} \qquad\qquad (7.9.1)$$

where $\phi_{i.}$, $\phi_{.j}$, and $\phi_{..}$ are the row, column, and overall means, respectively. Any measure of similarity, such as correlation, may be substituted for kinship. The two largest eigenvectors of the centroid-adjusted matrix give a two-dimensional plot in which similar populations are close together. Alternatively, the two-dimensional plot may be determined by a nonparametric method that considers only inequalities ($\phi_{ij} < \phi_{ik}$ or $> \phi_{ik}$) or by a least-squares method that maximizes the product-moment correlation between the centroid-adjusted matrix and its two-dimensional approximation. The latter is a reasonable measure of goodness-of-fit, but the precision of estimates of kinship is generally not the same for all pairs of populations. Since the method that maximizes the (unweighted) correlation is not necessarily the best, there is little basis for choice among methods. Most investigators use principal components (fig. 7.9.1).

Much differentiation is related to isolation by distance, and so it is helpful to translate and rotate any two-dimensional plot to maximize congruence with geographic coordinates, leaving the configuration invariant. Geographic distances are most conveniently measured in kilometers. If coordinates (x, y) are given in degrees of latitude and lon-

Hybridity   Lapp
                                               •
—— θ < 0.02
—–· θ < 0.04

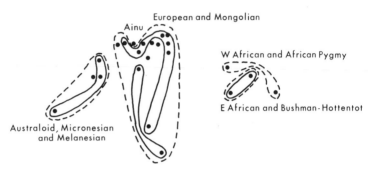

Fig. 7.9.1: Principal components of kinship among racial groups.

gitude, respectively, the coordinates in kilometers by plane trigonometry on the Hayford spheriod are

$$X = (X - \bar{x})(111.4175 \cos \beta - .0940 \cos 3\beta + .0002 \cos 5\beta) \qquad (7.9.2)$$

$$Y = (Y - \bar{y})(111.1363 - .5623 \cos 2\beta + .0011 \cos 4\beta)$$

where the mean coordinates are $\bar{x}$, $\bar{y}$ and $\beta = (y + \bar{y})/2$.

In the above topologies, similar populations tend to be located close together in a plane. There is another type of graph, called a "tree," in which dissimilarity is indicated by distance along one axis, the orthogonal axis serving merely to space the populations uniformly, so that the logical form of the tree is unchanged by rotation of any branch, as $(AB)(C) = (BA)(C) = (C)(AB) = (C)(BA)$. A phenetic tree, or *dendrogram*, is based entirely on phenotypic dissimilarity, which may be measured by hybridity θ_{ij} or alternatively by genetic distance,

$$D_{ij} \equiv \phi_{ii} + \phi_{ij} - 2\phi_{ij} \qquad (7.9.3)$$

At low levels of inbreeding θ_{ij} and D_{ij} differ only by a scalar. No phylogeny is implied by a dendrogram, although if the number of loci on which it is based is sufficiently large, the major branches (and less reliably, the minor branches) may have phylogenetic significance.

There are many ways to construct a dendrogram, which usually give slightly or markedly different trees. Although estimates of dissimilarity do not have the same precision for different pairs of populations, the correlation between dissimilarity and the estimate from the dendrogram gives a rough measure of adequacy of the tree. This correlation is often less than for kinship in planar topologies, reflecting the loss of information when a swarm of interbreeding populations is forced into a branching structure.

A phyletic tree, or *cladogram* (fig. 7.9.2), is an interpretation of a dendrogram in phylogenetic terms, and its meaningful axis represents time in years or generations. The time scale may be inferred from historical or paleontological evidence or, less accurately, by a transformation of hybridity, assuming a uniform divergence rate. A more realistic model founders on our ignorance of evolutionary sizes and systematic pressures during differentiation. We may suppose with glottochronology that

$$\phi_{ij} = ae^{-B(T-t_{ij})} \tag{7.9.4}$$

where a is the Malécot parameter for the first branch, T is the duration of the array as the time since the first split, and t_{ij} is the time from the origin of the array to differentiation of populations i and j (a, B > 0). This leads to

$$T - t_{ij} = T\left\{\frac{\ell n 2\theta^*/\phi_{ij}}{\ell n[(1-\theta^*)/\theta^*]}\right\} \quad \text{b.p.}, \tag{7.9.5}$$

where θ^* is the maximum hybridity in the array and "b.p." signifies "before present."

When two populations diverge, the decline of their kinship with time is likely to increase with increasing geographic and cultural distance, given exposure to different gene pools of migrants and perhaps to disruptive selection. This suggests that genetic divergence is not likely to be linearly exponential, but more nearly

$$\phi_{ij} = ae^{-B(T-t_{ij})^2} \tag{7.9.6}$$

which yields

$$T - t_{ij} = T\sqrt{\frac{\ell n(2\theta^*/\phi_{ij})}{\ell n[(1-\theta^*)/\theta^*]}} \quad \text{b.p.} \tag{7.9.7}$$

This equation has given results consistent with other evidence.

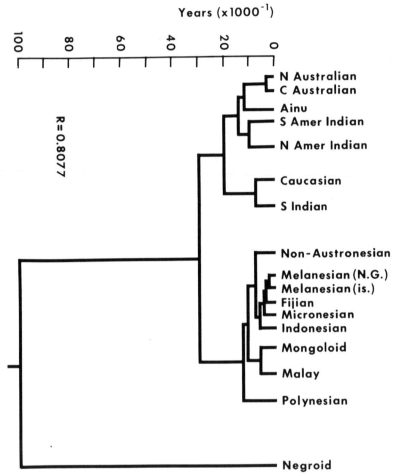

Fig. 7.9.2: A cladogram for racial groups.

Since trees permit branching but not anastomosis, and every human population undergoes "fusion" (ie, hybridization), a dendrogram of sub-specific differentiation cannot be in every branch a cladogram, even if the number of traits studied were sufficiently large that sampling error of hybridity could be neglected. The arbitrary but reasonable standard of numerical taxonomy that at least 100 traits be assayed is never approached by geneticists, and appreciable sampling errors must be assumed and have, in fact, been simulated.

The phenetic approach is almost universally employed in numerical taxonomy, which has always insisted that its trees are not cladograms.

However, interpretation of the major branches of a dendogram as a cladogram yields estimates of divergence times that may sometimes be interesting, although a population unusually differentiated because of small evolutionary size or systematic pressure tends to be assigned a spuriously high estimate of divergence time. Consistency with other evidence provides validation. The investigator interested in phylogeny should examine both a planar and tree representation. If he recognizes these topologies as a convenient distortion, he will be less likely to indulge in unwarranted phylogenetic speculation. Only the original kinship matrix contains all the information about differentiation, although in a form that often cannot be assimilated without reduction.

7.10 Questions

1. *Give equations 7.1.3 for unilineal relatives.*

   $P(aa \mid aa) = 4\phi q(1 - q) + q^2$

   $P(Aa \mid Aa) = 2\phi[1 - 4q(1 - q)] + 2q(1 - q)$

2. *Write equation 7.1.1 with exclusion of loops through a common ancestor to a more remote common ancestor, where F_i is the inbreeding coefficient of an admissible common ancestor.*

   $$\phi_{IJ} = \sum_{i=1}^{t} (\tfrac{1}{2})^{m_i + 1}(1 + F_i)$$

3. *An unfortunate investigator has a register of probands for some disease and a population genealogy, but does not know for certain whether pedigree members who are nonprobands are affected or not. What method could be used to test whether the frequency of affection is elevated among relatives of probands?*

   Equation 7.3.10 with ϕ_i in place of F_i.

4. *With probability ϕ_{ij}/ϕ_{ii} a gene in j is drawn from the same pool as a gene from i, and with probability $1 - \phi_{ij}/\phi_{ii}$ a gene in j is drawn from the same gene pool as a long-range migrant. Use this fact to estimate m_{ei}, N_{ei} from the transition matrix P and the matrix Φ of equilibrium kinship.*

   $m_{ei} \doteq m_i + (1 - m_i) \sum_{j=1}^{n} P_{ji}(1 - \phi_{ij}/\phi_{ii})$

   $N_{ei} \doteq (1 - \phi_{ii})/4\phi_{ii} m_{ei}$

5. *What is the main source of error in applying equation 7.3.10?*

Confounding of affection with inbreeding, for example by unequal pedigree depth for cases and controls, or environmental correlates of inbreeding (rurality, poverty, and so on), or by preferential consanguineous marriages of affected members. The latter may be avoided by restricting attention to offspring of normal × normal matings.

6. *What is the mean inbreeding of a region?*

$$\alpha = \Sigma N_i \alpha_i / \Sigma N_i$$

7. *Express the kinship of P, F_1, F_2 relative to an array of populations in terms of hybridity θ and F_2 kinship ϕ, and also in terms of ϕ_{ii}, ϕ_{jj} and ϕ_{ij}.*

$$P = \phi + (1 - \phi)\theta = \frac{\phi_{ii} + \phi_{jj}}{2}$$

$$F_1 = \phi - (1 - \phi)\theta = \phi_{ij}$$

$$F_2 = \phi = \frac{\phi_{ii} + \phi_{jj} + 2\phi_{ij}}{4}$$

8. *Verify that the above relation satisfies $F_2 = (P + F_1)/2$ and*

$$\theta = \frac{P - F_2}{1 - F_2} = -\frac{F_1 - F_2}{1 - F_2}.$$

9. *Express isolation by distance in terms of the hierarchical model.*

$$F_{ST} = ae^{-bd}$$

$$F_{RT} = \frac{-L}{1 - L}$$

$$F_{SR} = \frac{F_{ST} - F_{RT}}{1 - F_{RT}} = (1 - L)ae^{-bd} + L$$

10. *Express H_0, H_1, H_2, the heterozygosis in P, F_1, F_2, respectively, in terms of H, the heterozygosis under panmixia in the region from which the pair of populations is drawn and also in terms of H_2.*

$$H_2 = H(1 - \phi)$$

$$H_1 = H[1 - \{\phi - (1 - \phi)\theta\} = H_2(1 + \theta)$$

$$H_0 = H[1 - \{\phi + (1 - \phi)\theta\} = H_2(1 - \theta)$$

11. *Derive limiting kinship in an array of species so diverse that every allele is rare.*

$\Sigma Q^2 \to 0$

$$\phi_{ij} = E\left\{\frac{\sum_k q_{ki} q_{kj} - \Sigma Q_k^2}{1 - \Sigma Q_k^2}\right\} \to E\{\Sigma q_{ki} q_{kj}\}$$

12. *Does the above result have any utility with less diverse arrays?*

For a fixed set of alleles within a region it may be used topologically, but it cannot be predicted from genealogy or migration, provides no relation between gene and genotype frequencies, and varies erratically among loci and regions.

13. *Genetic topology has been taken as a problem in discrimination among populations. Discuss.*

The probability of correct assignment increases to unity in the limit as the number of independent traits increases. As a function of the number of traits, the probability of correct assignment is not a useful measure of topology.

14. *Derive Q in equation 7.7.3 for mates who are both heterozygotes for the same rare allele.*

From table 7.5.1 if $q > F$ the frequency of such matings is $4p^2q^2 + 4pq(1 - pq)F$, where $F = (1 - L)ae^{-bd} + L$. As $p \to 1$, $L \to 0$, this gives $Q = q + L$.

15. *Derive Q in equation 7.7.3 for mates who are both homozygotes for the same rare allele.*

$Q \doteq q/6 + L$

16. *What is the difference between table 7.8.1 and expectations in term of hybridity?*

Table 7.8.1 relates to mean gene frequency in the pair of populations. Hybridity is related to the gene frequencies in the array from which the pair of populations was drawn.

17. *Does an estimate of small Q from equation 7.7.3 prove a genetic etiology?*

   No, since environmental causes may also cluster in space. However, verification of a predicted value of Q, or two significantly different estimates of Q, may support a genetic interpretation.

18. *Migration data predicting inbreeding give good estimates of m_e, but N_e is often unreliable. It has been found that m_e is stable over a wide range of assumed N_e. Suggest a way to get good estimates of both parameters.*

   A. Estimate m_e from migration at trial values of N.
   B. At this value of m_e, estimate N_e from genealogies.
   C. Adjust N to CN, where C is the ratio of initial estimates of N_e from migration and genealogy, and verify that migration then gives essentially the same estimates of N_e and $\phi^{(\infty)}$ as genealogy.

19. *Suggest a definition of stepping-stone distance D as a function of the short-range migration rate k and the standard deviation of short-range migration.*

   $D \equiv \sigma'/\sqrt{k}$ corresponds to $(\sigma')^2 = kD^2$.

20. *From the χ^2 formula for a codominant system, suggest an estimate for kinship where gene frequencies are known, but not the phenotype frequencies from which they were calculated.*

$$\phi_{ij} = \sum_{k-1}^{m} \left\{ \frac{q_{ik}\,q_{jk}}{Q_k} - 1 \right\} \Big/ (m-1)$$

where

$$q_{ik}^2 = \frac{n_{ik}(n_{ik}-1)}{2N_i(2N_i-1)}$$

$$q_{ik}\,q_{jk} = \frac{n_{ik}\,n_{jk}}{4N_i^2}$$

$$n_{ik} = 2q_{ik}\,N_i$$

$$Q_k = \Sigma n_{ik}/2\Sigma N_i$$

21. *A matrix of population similarities can be constructed in indefinitely many ways, for example from classification probabilities in discriminant analysis and from the sign of kinship in multiple kinship matrices. What are the advantages and disadvantages of such measures of population structure?*

Like kinship they give a topology for populations within a region. Unlike kinship they give no meaningful comparison among regions and no precise genetic interpretation.

22. *It has been suggested that distance be measured not linearly but as people actually travel. What are the advantages and disadvantages of this approach?*

It reduces the effect of geographic and political barriers. However, we usually cannot specify how people migrate for all $n(n-1)/2$ pairs of populations in a migration matrix. Any attempt to approximate this, for example by the shortest road distance, is arbitrary and imprecise and exposes the measurement of distance to erratic changes with activity of the highway department. There is no reason to think that the probability of migration from one point to another, depending on economic and personal factors, is related in any simple way to the shortest road distance, or that all types of roads from footpaths to highways are equally traveled.

23. *Show that expected homozygosity equals kinship if there are many alleles with the same expected frequency.*

ΣQ^2 is negligible by comparison with $\Sigma Q = 1$, and so

$$\Sigma Q^2(1 - \phi) + \Sigma Q\phi \to \phi$$

24. *Under a polygenic model affection p_i declines roughly exponentially with kinship to a proband. Derive a maximum likelihood test of the null hypothesis of no familial aggregation, using A_i for the number of affected and C_i for the number of normals among relatives with kinship ϕ_i to a proband.*

$$p_i = pe^{b\phi_i}, \qquad r = p/(1 - p)$$

$$U_i \equiv \left.\frac{\partial \ell nL}{\partial b}\right|_{b=0} = \left.\frac{\partial \ell nL}{\partial p_i}\left(\frac{\partial p_i}{\partial b}\right)\right|_{b=0}$$

$$= (A_i - rC_i)\phi_i$$

$$K_i = \left.\frac{-\partial^2 \ell nL}{\partial p_i^2}\left(\frac{\partial p_i}{\partial b}\right)^2\right|_{b=0} = r(A_i + C_i)\phi_i^2$$

test of H_0 $(b = 0)$: $\chi_1^2 = (\Sigma U_i)^2/\Sigma K_i$
test of homogeneity: $\chi_{n-1}^2 = \Sigma(U_i^2/K_i) - (\Sigma U_i)^2/\Sigma K_i$

25. *Why is the above test not in common use?*

All models for family resemblance give a test of the null hypothesis. Segregation analysis is more powerful and reliable than a test on kinship, which assumes that relatives are independent.

7.11 Bibliography

Crow JF, Denniston C: Genetic Distance. Plenum Press, New York, 1974.

Jacquard A: Genetics of Human Populations. Freeman, Cooper and Co., San Francisco, 1978

Malécot G: The Mathematics of Heredity. Freeman, Cooper and Co., San Francisco, 1969

Morton N.E.(ed): Genetic Structure of Populations. University Press of Hawaii, Honolulu, 1973

Roberts DF, Sunderland E: Genetic Variation in Britain. Taylor and Francis Ltd, London, 1973

Weiner JS, Huizinga J: Assessment of Population Affinities in Man. Clarendon Press, Oxford, 1972

Wright S: Evolution and the Genetics of Populations. II. The Theory of Gene Frequencies. University of Chicago Press, Chicago, 1969

8. Frequency, Fitness and Mutation

The contemporary distribution of disease reflects opposing pressures of selection and mutation. We shall now consider how these pressures are estimated and their effect on disease frequency.

8.1 Frequency

In genetics and in common usage the terms frequency, prevalence, and incidence are synonymous. In genetic epidemiology *prevalence* (P) is the frequency of a genotype or phenotype in the general population at a particular time. Lifetime *incidence* (I) refers to the frequency of a genotype or phenotype among a cohort of conceptions or births. Thus *frequency* subsumes both prevalence and incidence. If there is delayed onset or selective mortality, incidence is greater than the corresponding prevalence $(I > P)$.

Consider some category k, perhaps defined on age and sex. Let g_k be the frequency of that category in the general population and m_k be the specific risk for affection. Then the prevalence of affection is

$$P = \sum_k g_k m_k,$$

(8.1.1)

which differs from the morbid risk defined in equation 4.1.5 by being based on the general population instead of a subpopulation.

For a common trait, it is clearly desirable to estimate prevalence from a random sample of the population. However, this is prohibitively expensive for a rare disease, which is ordinarily ascertained through probands. Then equation 4.1.4 estimates the number of affected individuals in the population as $R = A/\pi$, where π is the ascertainment probability and A is the number of probands. If N is the corresponding population size, prevalence is given by

$$P = R/N$$

(8.1.2)

The investigator must be fastidious about defining π, A, and N in time, space, and condition. If an affected person must be alive to be a proband, N is the number of individuals living at the time of investigation in the defined catchment area from which probands were drawn. If dead individuals may be probands, N is the number of births in the defined catchment area over a suitable time interval. Migration affects the catchment area, which in general may consist of regions with different ascertainment probabilities. If π_i is the ascertainment probability in the ith stratum, with A_i probands and size N_i, then

$$P = \frac{\Sigma A_i}{\Sigma \pi_i N_i} \tag{8.1.3}$$

Although prevalence usually relates to a phenotype, occasionally it is applied to the frequency of individuals who would be affected if they survived to a ripe age (say, 70 years), or to the frequency of a genotype in the general population, concepts which are more conveniently dealt with in terms of incidence.

Let p_k be the probability that an ultimately affected individual be affected before he has passed the midpoint of k and not die of affection before then. The incidence, at conception, of individuals who will ultimately be affected is defined on prevalence P as

$$I = P / \sum_k g_k p_k \tag{8.1.4}$$

If ages of onset and death of probands are typical they permit an estimate of the p_k, otherwise a random sample of affected must be taken. If affection is entirely due to a major locus, which is invariably expressed before age 70, then equation 8.1.4 also gives the genotypic incidence at conception I_G. In general,

$$I_G = I/f \tag{8.1.5}$$

where f is the probability of affection among individuals with the major locus genotype who are either affected or survive to a ripe age without affection.

Segregation analysis with the mixed model or the generalized single locus model (table 4.3.1) gives an estimate of gene frequency q from which genotypic incidence may be estimated as

$$I_G = \begin{cases} q^2 \text{ for a recessive gene under panmixia} \\ 2q(1-q) + q^2 \text{ for a dominant gene} \\ q \text{ for a sex-linked recessive in males} \end{cases} \tag{8.1.6}$$

This is usually interpreted as the incidence at conception, assuming no differential mortality before diagnosis.

8.2 Fitness

There are three components to fitness: fertility, survival, and generation time. The *relative fitness* of the kth genotype or phenotype is

$$W_k = B_k M_k / T_k \tag{8.2.1}$$

where

B_k = ratio of progeny per reproductively mature individual in k and in the population

M_k = ratio of survival to reproductive maturity in k and in the population

T_k = ratio of the mean age at reproduction of individuals in k and in the population

If selection is largely through mortality, individuals should be enumerated as soon after conception as practicable. If selection is largely through reduced fertility, individuals may be enumerated at reproductive maturity. Children should be enumerated at the same age as parents. In practice, a control drawn from pedigree members is substituted for the population.

Fitness is normal if $W_k = 1$. The selection coefficient in the kth genotype or phenotype is

$$m_k = 1 - W_k \tag{8.2.2}$$

Absolute fitness of the kth genotype or phenotype is

$$F_k = RW_k \tag{8.2.3}$$

where R is half the mean number of children per individual in the general population.

If the segregation parameter p differs from p_0 only because of mortality before the age at observation, and subsequent selective mortality is negligible, then

$$p = \frac{Mp_0}{Mp_0 + 1 - p_0} \tag{8.2.4}$$

which inverts to

$$M = \frac{p(1 - p_0)}{p_0(1 - p)} \tag{8.2.5}$$

$$= p/(1 - p) \text{ for a rare dominant } (p_0 = \tfrac{1}{2})$$

$$= 3p/(1 - p) \text{ for a rare recessive } (p_0 = \tfrac{1}{4})$$

An alternative calculation regresses in a sample of trait-bearers and controls the number of children S on relevant variables (not including carrier status) to obtain first a predicted number $E(S) > 0$ and then the weighted regression

$$y = a + cX$$

where $y = S/E(S)$, and X is trait-bearer status as a binary (0, 1) variable. The weight appropriate to a Poisson distribution within a cohort characterized by the same value of the predictor is $E(S)$. The estimate of relative survival is

$$M = 1 + c/a \tag{8.2.6}$$

A similar calculation estimates B from $E(B)$ or the relative fetal death rate D from $E(D)$. For relative generation time, age at conception $E(A)$ is estimated with equal weights, since it is not a Poisson variable. Matched controls would permit a simpler calculation, but selection of pedigrees through probands virtually compels use of normal pedigree members as controls and, therefore, use of equation 8.2.6 to remove ascertainment bias.

Assuming that B, T, M are independent, the large sample variance of an estimate of W from equation 8.2.1 is

$$\sigma_W^2 = W^2(\sigma_B^2/B^2 + \sigma_T^2/T^2 + \sigma_M^2/M^2) \tag{8.2.7}$$

and the quantity

$$\chi_1^2 = \frac{(W - 1)^2}{\sigma_B^2 + \sigma_T^2 + \sigma_M^2} \tag{8.2.8}$$

tests the null hypothesis of normal fitness. If any of the values B, T, M is taken as unity, the corresponding variance is omitted.

8.3 Autosomal Dominants

The mutation rate for deleterious autosomal dominants may be estimated by three methods—direct, semidirect, and indirect.

If a proportion x of probands have normal parents, if penetrance is complete, if somatic mutation, phenocopies, germinal mosaicism, unrecognized recessive genes, and selective mortality before diagnosis are negligible, and if parentage error is reasonably excluded, the *direct* estimate of mutation rate is

$$u = xI/2 \qquad (8.3.1)$$

per gamete per generation, where I is the incidence of carriers.

If all these assumptions hold except the penetrance is incomplete, we may estimate by segregation analysis the proportion x of cases that are sporadic (eq. 5.4.1). Then if the segregation frequency differs from 1/2 only because of incomplete penetrance, the expected value is

$$p = f/2 \qquad (8.3.2)$$

which inverts to

$$f = 2p \qquad (8.3.3)$$

where f is the penetrance. The *semidirect* estimate of the mutation rate is formally the same as equation 8.3.1, with x estimated by segregation analysis instead of being directly observed and I taken as the genotypic incidence in equation 8.1.5. The mixed model of segregation analysis estimates x and the gene frequency q, assuming a single locus. As $q \to 0$,

$$I = 2q(1 - q) + q^2 \to 2q \qquad (8.3.4)$$

which accounts for the factor of 2 in equation 8.3.1. If there is more than one locus, $I \to 2\Sigma q$.

The indirect method assumes that relative fitness and mutation rate are constant in a given population over many generations, during which an equilibrium between selection and mutation has been reached. Then if m is the selection coefficient against carriers estimated from relative fitness by equation 8.2.2, the *indirect* estimate of mutation rate is

$$\mu = mI/2 \quad \text{per gamete per generation.} \qquad (8.3.5)$$

This depends on the principle that at equilibrium the gain of genes by mutation (xI) balances loss by selection (mI), or

$$m = x \qquad\qquad (8.3.6)$$

The mean persistence of a mutant is $1/m$ generations.

Agreement among different methods of estimation increases confidence in the mutation rate, but this validation is imperfect because the critical assumptions are the same for all methods (table 8.3.1). With small families, germinal mosaicism cannot be excluded. Incomplete penetrance and differential mortality before diagnosis are negligible if the segregation frequency p in a good body of data is close to $1/2$, or equivalently if displacement in the mixed model is large. Phenocopies (including somatic mutation) and unrecognized recessive cases can be excluded indirectly if segregation analysis under equation 5.2.2 suggests that possible mutants can segregate ($h = 0$), but this evidence requires that affected children of normal parents be fertile and that complete selection be applied to children of possible mutants. An isolated case due to paternity error can reasonably be excluded for a disease of low fitness, especially if other genetic systems are consistent with the nominal father.

Table 8.3.1: Estimates of dominant mutation rate for ostensibly single loci in man

Disease	Mutation rate per locus $u \times 10^6$
Facioscapulohumeral muscular dystrophy	1
Achondroplasia	10
Aniridia	3
Dystrophia myotonica	10
Retinoblastoma	8
Acrocephalosyndactyly	4
Osteogenesis imperfecta	10
Tuberous sclerosis	8
Neurofibromatosis	73
Intestinal polyposis	13
Marfan's syndrome	5
Polycystic disease of kidneys	92
Multiple exostosis	8
von Hippel-Lindau	1
Pelger anomaly	6
Spherocytosis	22
Microphthalmos	6
Waardenberg's syndrome	4
Nail-patella syndrome	2
Huntington's chorea	2
Multiple telangiectasia	2

A serious limitation of dominant mutation rates is that they are expressed per gamete, usually without any means to determine the number of contributory loci. A second problem is that systematic study to determine incidence and segregation parameters is not feasible for extremely rare diseases, and so there is bias toward estimation of high mutation rates. Attempted corrections depend on untested assumptions about the distribution of mutation rates among loci.

Matings involving rare homozygotes, multiple alleles, or linkage could, in principle, identify the parent in whom a mutation occurred, but this is seldom feasible. A few dominant diseases, notably achondroplasia, acrocephalosyndactyly, and Marfan's syndrome, show elevated paternal age at conception of apparent mutants, indicating that some mutational events increase with age of spermatogonia or duration of the haplophase.

An appreciable fraction of spontaneous abortions may be due to dominant mutations, but no method to estimate their frequency has been proposed.

8.4 Sex-linked Recessives

The direct estimate of mutation rate is not applicable to sex-linked recessives, since an isolated proband may be a chance-isolated segregant from a carrier mother, who by definition of recessivity cannot be recognized with certainty. The semidirect estimate can be made from x and p of equation 5.4.1 or from x and $I = q$ by the mixed model with pointers. The indirect method is also feasible, if the ratio of mutation rates in egg and sperm is known.

The equilibrium condition was determined by Haldane as

$$x = \frac{mu}{2u + v}, \qquad u = xI \qquad\qquad (8.4.1)$$

where m is the selection coefficient against affected males and u, v are the mutation rates in egg and sperm, respectively. Deviation of x from $m/3$ therefore provides a test of the equality of u and v, assuming mutation − selection equilibrium. As discussed in 5.1, an autosomal sex-linked gene has $x = m/(m + 1)$, and so for a nearly lethal gene a frequency of sporadic cases significantly less than $1/2$ is evidence of sex-linkage.

The brothers of a carrier woman are not at risk if she received the gene by mutation or from an affected father. The probability that the brothers of a random carrier not be at risk is $x' = 1/2$ (table 8.4.1). The complement $1 − x'$ is the probability that the mother of a random carrier

Table 8.4.1: The probability (x') that the brothers of a random carrier not be at risk

Sex-linked disease	Estimate x'	ML scores at x' = 1/2		
		$U_{x'}$	$K_{x'x'}$	χ^2
Duchenne muscular dystrophy	0.59	18.799	205.69	1.720
Hemophilia (A + B)	0.48	− 5.953	331.40	0.107
Lesch-Nyhan	0.56	5.520	98.00	0.311
Total	0.52	18.366	635.09	0.531

be a carrier. Although this result depends on mutation − selection equilibrium, x' does not depend on selection or mutation rates and therefore provides a valuable check on the quality of segregation data. The theory is satisfied for the three sex-linked diseases studied.

Duchenne muscular dystrophy is characterized by (a) sex-linkage; (b) onset usually in the first three years of life, but occasionally as late as the third decade; (c) symmetrical involvement, first of the pelvic girdle musculature, and later of the shoulder girdles; (d) pseudohypertrophy of calf muscles; (e) absence of abortive or partially affected cases; (f) steady and rapid progression usually leading to inability to walk within 10 years of the onset; (g) progressive deformity with muscular contractures, skeletal distortion, and atrophy; and (h) death from inanition, respiratory infection, or cardiac failure usually in the second decade, but sometimes not until middle life. Affected individuals rarely reproduce. Differential diagnosis requires exclusion of limb girdle muscular dystrophy, a mixture of recessive cases and phenocopies which in some families manifest pseudohypertrophy but tend to be less severe with later age of onset, confinement to bed, and death, and less constant development of pseudohypertrophy and contractures. Absence of affected females, nonconsanguineous parentage, and presence of other affected males related through normal females, in conjunction with the typical symptomatology, favor the diagnosis of Duchenne muscular dystrophy.

Most Duchenne cases are of the classical severe type, but a minority are much milder. Some neurologists emphasize the distinction between the two groups, the severe form having onset before the age of 10, with death usually by age 20, while the mild form (Becker) resembles limb girdle muscular dystrophy in commonly having onset after the age of 10, with patients able to walk in maturity, and death occurring after age 25, sometimes in old age. There is some overlap, and pedigrees have been described in which both forms occur. Whether the Becker type is caused by a different allele, a different sex-linked locus, or by modifying factors has not been determined. It may not be genetically homogeneous, although linkage data suggest a locus on Xq. Since the Becker type is much rarer than typical Duchenne cases, the undoubted variation in diagnostic standards has little effect on estimates of gene frequency and mutation rate for Duchenne muscular dystrophy (table 8.4.2).

The mutation rate of 8.8×10^{-5} per locus per generation is unusually high. Perhaps the unknown product of the *MDD* locus has a critical target of unusual size: ie, alteration in one of many codons can produce the disease. Alternatively, there may be a critical target of normal size but unusual mutability. Affected females are rarely observed. About half of these are XO hemizygotes, providing conclusive evidence for sex-linkage. The remainder are X-autosome translocations in which the normal chromosome is preferentially inactivated. All of these translocations have the break in band P21 of the X chromosome, strongly suggesting that the locus for *MDD* lies in that band. Only one other X/A translocation has been reported as a probable cause for an affected female (a single instance of anhidrotic ectodermal dysplasia). Compared to other sex-linked diseases, Duchenne cases are in as much excess among X/A translocations as among affected males. Unfortunately, this does not conclusively favor the mutable site or large target hypothesis.

In addition to rare XO and X/A females, XY females with testicular feminization are predicted to show typical expression if they carry the MDD gene. XX carriers occasionally show milder symptoms, including kyphosis, enlarged calves, and muscle wasting, sufficient to support a diagnosis of limb girdle muscular dystrophy. The majority of carriers can be recognized through elevated serum levels of enzymes leaked from muscle. Creatine kinase is the most reliable, although transaminase, aldolase and other enzymes are also useful. A low frequency of lymphocyte capping has given good discrimination in some laboratories but not others. Many other carrier tests have been proposed, including electromyography, circulation time, and erythrocyte spectrin peptides, which

Table 8.4.2: Estimates of gene frequency (q) and mutation rate (u) per million male births

Sex-linked disease	$q \times 10^6$	$u \times 10^6$
Duchenne muscular dystrophy	265.0	88.00
Hemophilia A	55.0	13.10
Hemophilia B	6.0	0.55
Lesch-Nyhan	5.2	1.70

fail when repeated in systematic blind studies. At the present time, no combination of such tests gives perfect carrier diagnosis, and the probability that a woman is a carrier must be used in genetic counseling. The semidirect estimate of mutation rate is consistent with equality in sperm and egg.

Hemophilia has the advantage over Duchenne muscular dystrophy that its gene product is well characterized, leading to recognition of two sex-linked loci: classical hemophilia A (*HEMA*) due to deficiency of coagulation factor VIII, and hemophilia B (*HEMB*) or Christmas disease due to deficiency of coagulation factor IX. The two loci are not closely linked, and their products are structurally different. Carrier tests give useful but imperfect discrimination.

Compared with Duchenne muscular dystrophy, hemophilia has the disadvantages for mutation research of selective mortality before diagnosis and of departure from mutation-selection equilibrium. When hemophilia causes death in infancy, the diagnosis may not be made. An isolated case is less likely to be diagnosed. Recent advances in therapy and inclusion of mild cases have reduced the selection coefficient, which in the past must have approached unity, to an extent that can only be guessed. It will take at least several generations to reach a new equilibrium, perhaps at m = .25 or less. With increasing availability of factor VIII, m and x may gradually decrease over many generations. Despite these deviations from equilibrium, the frequency of sporadic cases has been in satisfactory agreement with expectation under equal mutation rates (table 8.4.3).

Lesch-Nyhan disease is characterized by mental retardation, spastic cerebral palsy, choreoathetosis, uric acid urinary stones, and self-destructive biting of fingers and lips. It is caused by deficiency of the enzyme hypoxanthine guanine phosphoribosyl transferase (HGPRT), the locus for which is on the X chromosome. Whereas Duchenne muscular

Table 8.4.3: Estimates of m and x

Sex-linked disease	Selection coefficient (m)	Frequency of sporadic cases (x)		
		Predicted if u = v	Observed	χ_1^2
Duchenne muscular dystrophy	0.962	0.32	0.35	0.51
Hemophilia (A + B)	0.750	0.25	0.25	0.00
	0.620	0.21	0.18	0.07
Lesch-Nyhan	1.000	0.33	0.07	8.85

dystrophy and hemophilia typify carrier tests with substantial error, Lesch-Nyhan disease approaches the ideal of perfect carrier detection, at least with cloning and hair follicle analysis. Cell selection and auto-radiography have been less reliable. Imperfect carrier diagnosis leads to tedious calculations and painful risk taking. With perfect diagnosis, these practical problems are eliminated.

Unfortunately, sampling problems are severe for mutation research on Lesch-Nyhan disease, which is too rare to have been the subject of systematic epidemiological study. Patients must be referred by clinicians to specialized laboratories. Patients with affected relatives are more likely to attract interest under two conditions: if the mode of inheritance is not fully established, or if clinical diagnosis is difficult. Compulsive self-destructive behavior is the most striking, but inconsistent, presenting symptom. Death during the first decade is common, and many affected, especially isolated cases, must die undiagnosed. Clearly the probability of at least one living patient with pathognomonic symptoms increases with the number of affected in a pedigree.

For Lesch-Nyhan disease, unlike other sex-linked conditions, there is a significant deficiency of sporadic cases. We are left with a logical dis-junction: either Lesch-Nyhan disease has a higher mutation rate in males than in females—although Duchenne muscular dystrophy does not (in which case each sex-linked disease must be considered separately), or there is a bias toward referring familial Lesch-Nyhan cases to specialized centers dealing with inherited diseases. Only the combination of sound epidemiological surveys with impeccable laboratory tests will ultimately resolve this question.

An indirect approach can be made through increase in mutation with age, but there are many obstacles. The sex with the higher mutation rate (if there is a difference) may show a smaller age effect, and age trends are unpredictably curvilinear. Although several dominant mutation rates increase with age in males, we do not know how many of these heter-ozygous mutations are deletions and frameshifts that would be inviable if hemizygous. The possibility that sex-linked mutations increase with age in males has been suggested by two studies and denied by others. The problem of a suitable control is delicate for sex-linked genes because the probability of ascertaining mutation increases with the number of grand-children at risk, so there is selection for noncontraceptive couples. A conservative approach uses the age distribution of paternal grandfathers as control for maternal grandfathers, whose mutations are being assayed. This test has so far been negative.

The latest data from Drosophila show no sex difference in mutation rate for autosomal and sex-linked lethals: earlier reports had been conflicting. In the mouse the frequency of spontaneous mutations is higher in males, but the difference does not approach significance (table 8.4.4).

8.5 Autosomal Recessives

Only the indirect estimate of mutation rate is applicable to rare autosomal recessives in man, since mutation usually occurs many generations before manifestation in a homozygote (table 8.5.1). The equilibrium condition for a complete recessive under panmixia is

$$u = mI \tag{8.5.1}$$

If q is estimated from segregation analysis in a population with mean inbreeding α, the estimate will be inflated to satisfy approximately

$$I = q^2 = q'(q' + \alpha) \tag{8.5.2}$$

where q' is the true gene frequency. If there is more than one locus, the term on the right is replaced by $A' + B'\alpha$ where A', B' are the genetic loads defined in equation 7.3.4.

Estimates of recessive mutation rates are free of some of the assumptions for dominant and sex-linked mutation. Germinal mosaicism and parentage error have no appreciable effect on the indirect estimate. Incomplete penetrance, somatic mutation, and phenocopies are controlled by segregation analysis. Problems of multiple loci and biased ascertainment of higher rates remain. The assumption of mutation-selection equilibrium is more restrictive because effects of heterozygosity on fitness are immeasurable but perhaps not negligible, and approach to equilibrium is exceedingly slow when mutation, selection, or inbreeding is altered. The incentive to estimate mutation rates from recessive genes comes from evidence that many detrimental mutations are recessive, and so the

Table 8.4.4: Spontaneous specific-locus mutations from wild type in the mouse

Sex	Loci tested	Mutations	Rate per generation
Male	4,669,203	36	7.7×10^{-6}
Female	1,419,684	7	4.9×10^{-6}
Total	6,088,887	43 $\chi_1^2 = 1.19$	7.1×10^{-6}

Table 8.5.1: Recessive lethal mutation in Drosophila melanogaster

Spontaneous lethals/X chromosome	$1586/617,769 = .00257$
Ratio of salivary bands in chromosome II and X	$1939/1011 = 1.92$
Ratio of physical length II/X	$816/414 = 1.97$
Extrapolated spontaneous lethals/chromosome II	.0050
Observed spontaneous lethals/chromosome II	$294/58,496 = .0050$
Lethal load for chromosome II	.247
Mean persistence for lethals	$.247/.0050 = 49$ generations
Mean of $(F + q + h)m$ for lethals	$.0050/.247 = .020$
Number of loci on II as salivary bands	1939
Lethal mutations/locus/generation from gametic rate	$.0050/1939 = 2.6 \times 10^{-6}$
Mean lethal gene frequency per locus	$.247/1939 = 1.3 \times 10^{-4}$
Induced lethals/male X/acute R to sperm	3.0×10^{-5}
Doubling dose for acute radiation of sperm	$.00257/(3.0 \times 10^{-5}) = 86$ R

handful of dominant and sex-linked mutation rates that have been estimated in man are atypical (table 8.5.2).

If fitness is reduced in heterozygotes, we may use the notation of
table 7.3.1 with m substituted for s to denote selection rather than affection. Then the probability of selective elimination of a particular detri-

Table 8.5.2: Estimates of autosomal recessive mutation rate for ostensibly single
loci in man

Disease	Gene frequency (q)	Selection coefficient (m)	Mutation rate per locus (u $\times 10^6$)
Albinism		.50	
Caucasian	.0026		13
Japanese	.0052		26
Infantile amaurotic idiocy		1.00	
Caucasian	.0012		12
Japanese	.0020		20
Ichthyosis congenita		1.00	
Caucasian	.0012		12
Japanese	.0012		12
Congenital total color blindness		.50	
Caucasian	.0040		20
Japanese	.0055		28
Amyotonia congenita	.0020	1.00	20
Phenylketonuria	.0025	1.00	25
Microcephaly			
Caucasian	.0052	1.00	52
Japanese	.0049	.98	48
Microphthalmos	.0015	1.00	15
Epidermolysis bullosa dystrophica	.0036	.60	22
Spinocerebellar ataxia	.0025	.40	10
Cerebellar ataxia	.0005	.37	2
Cystic fibrosis (noncaucasians)	.0033	1.00	33

mental gene is Fm through inbreeding, (1-F)qm through random homozygosity, and (1-F)(1-q)hm through heterozygosity. The total probability of elimination is $1/\rho$, where the *persistence* ρ is the mean number of generations a particular detrimental gene survives before selective elimination (fig. 8.5.1). We deduce that

$$1/\rho = [F + (1 - F)q + (1 - F)(1 - q)h]m \qquad (8.5.4)$$

Since $1 - F$ and $1 - q$ are close to unity, this simplifies to

$$1/\rho \doteq (F + q + h)m \qquad (8.5.5)$$

At mutation-selection equilibrium the quantities F, q, h take the values under which equilibrium was approached, and the gene frequency equals the product of the mutation rate and persistence, or

$$u = q/\rho \doteq q(F + q + h)m \qquad (8.5.6)$$

This theory depends on two critical assumptions: the load is mutational

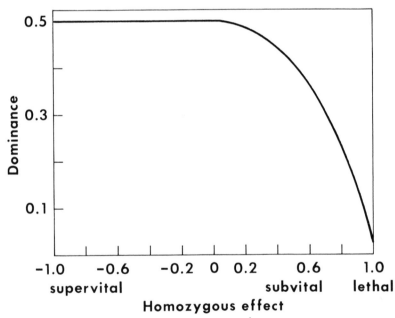

Fig. 8.5.1: Dominance (h) and homozygous effect (s) for viability in Drosophila based on partially heterozygous chromosome II.

Table 8.5.3: Application of genetic load theory to Drosophila mortality

L/gamete	.667
D/gamete	.637
Total mutation rate/gamete	.127(L + D) = .16561
Total mutation rate/locus	.16561/5149 = 32.2 × 10⁻⁶

$(h \geq 0)$ and F, q, h may be evaluated at equilibrium values, which in general are not contemporary ones. Since hm is too small to be directly measured, the only way to test the first assumption is by agreement of mutation rate estimates from equation 8.5.6 with other evidence. Departure of the parameters F, q, h from equilibrium in contemporary populations may be significant only for F.

The mean inbreeding in contemporary isolates has been estimated by equation 7.4.4 as .01 in populations that do not practice preferential consanguineous marriage and .02 in Middle Eastern populations in which close consanguineous marriage is preferred. National populations have smaller values of inbreeding. The value of inbreeding in the past when equilibrium was approached is conjectural. To the extent that local populations were smaller and more isolated, inbreeding may have been greater. However, populations that are increasing in size have more opportunity for consanguineous marriage. For example, the number of first cousins in a population with exactly C surviving children per family is $2C(C - 1)$, which increases more than linearly with C. Thus there is reason to speculate that inbreeding was less in the past than in current isolates, most of which have undergone recent expansion and are in any case selected for study because of conspicuous inbreeding. Roman Cath-

Table 8.5.4: Spontaneous mutation rate estimates per locus per generation × 10⁶

Source	Drastic (u_L)	Detrimental (u_D)	Total deleterious (u_T)
Single loci: recessive	12	10	22
dominant	4	10	14
sex-linked	3	17	20
Genotypic incidence	6	11	17
Phenotypic incidence	8	17	25
Private electrophoretic variants	7	14	21
Drosophila	5	27	32
Mouse	7	14	21
Mean	6	15	21

olic dispensations for consanguineous marriage do not indicate any consistent change in inbreeding during the several centuries before 1900 when an increase in effective migration rate was compensated by a decrease in evolutionary size. Inbreeding at equilibrium appears to be substantially greater than the gene frequency for most recessive diseases.

There has been a controversy about how variability in mutation rate estimates should be interpreted. One approach fixes attention on a specific phenotype and supposes that the probability of estimating a mutation rate is proportional to its value. Then the harmonic mean of such estimates is unbiased. The alternative approach takes as its objective the total deleterious mutation rate, assumed fairly uniform among loci, while the proportion of deleterious mutations that produces a particular phenotype is much more variable. Then the unbiased estimate of the deleterious mutation rate is the arithmetical mean. Although the two models are somewhat artificial, it seems reasonable to use the harmonic mean for a specific phenotype (including drastic mutations) and the arithmetical mean for the total deleterious mutation rate.

A different principle must be invoked to estimate the total deleterious mutation rate from the inbred load B. At equilibrium the mean of $F + q + h$ has been estimated as .01 for drastic genes, .25 for minor detrimentals, and .127 for all deleterious genes. If the affection probability s is equal to the selection coefficient m, then the mutation rate per gamete is .127B, about one-third of which is due to drastic genes.

8.6 Induced Mutation

There is a striking relation between radiosensitivity and quantity of DNA per haploid genome. How can complexity of a locus, as reflected by its target size, increase with total DNA content, when there is no comparable increase in the molecular weight of the protein product? This paradox could be explained by base-pair deletions or insertions in intervening sequences, which are not transcribed but can mutate to transcribable frameshifts that alter adjacent gene products. From DNA content the interpolated mutation rate in man is 2.6×10^{-7} serious detrimental mutations per locus per rad of acute gamma radiation. This value is important for the study of mutation in man, since it leads to estimates of doubling dose and number of loci, and thereby to a test of consistency for the spontaneous mutation rate (table 8.6.1).

The doubling dose D of a mutagen is the quantity required to double the spontaneous mutation rate u. If the frequency of mutation is

Table 8.6.1: Mutation parameters in various organisms. Radiation is assumed given to Drosophilia sperm and to mammalian spermatogonia or primary oocytes.

Organism	DNA per haploid genome (pg)	Specific locus mutations/acute r	Spontaneous mutations/locus /generation	Acute doubling dose (r)	Months/ generation
D. melanogaster	.20	3.0×10^{-8}	2.6×10^{-6}	86	1
M. musculus	2.26	1.7×10^{-7}	7.1×10^{-6}	42	3
H. sapiens	2.9	2.6×10^{-7}	1×10^{-5}	38	360

linear on dose,

$$D = u/v \qquad (8.6.1)$$

where v is the induced rate per unit of mutagen. In principle each type of mutation has its own doubling dose, although for the mouse there is good agreement among estimates of D for translocations, dominant and recessive visibles, and recessive lethals.

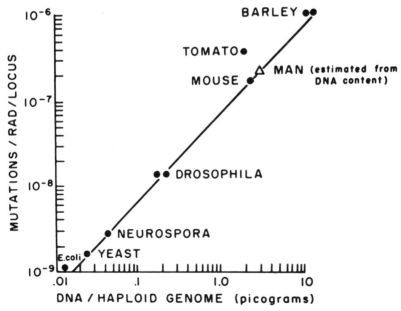

Fig. 8.6.1: Relation between forward mutation rate per locus per rad and the DNA content per haploid genome.

The effective number of loci per gamete is

$$n = U/u \tag{8.6.2}$$

where U is the mutation rate to serious detrimentals per gamete and u is the rate per randomly selected locus. Mammals give about 10^4 for n.

An independent calculation applies v from radiosensitity to the sex ratio at birth. Let p be the frequency of males in a control sample, and p_d be the frequency when the mother has received a mutagenic dose d, which induces Kd lethal mutations per X chromosome. Then

$$p_d = \frac{pe^{-Kd}}{1 - p + pe^{-Kd}} \tag{8.6.3}$$

If nearly all serious detrimentals among induced mutations cause fetal death, the number of loci per gamete is

$$n = K/vg \tag{8.6.4}$$

where g, the proportion of loci on the X chromosome, may be estimated from physical lengths of chromosomes. Consistency of these estimates for mammals supports the genetic load theory on which estimates of spontaneous mutation rates are based (table 8.6.2).

For chemical mutagens the response rate is usually not linear on dose. Although equation 8.6.1 cannot be used, the doubling dose exists as the solution of

$$f(d) = F(d) - F(0) = u \tag{8.6.5}$$

where f(d) is the specific response to a dose d, F(d) is the crude response with control F(0), and u is the spontaneous mutation rate. Since $u = F(0)$, this reduces to

$$F(d) = 2F(0) \tag{8.6.6}$$

Determination of a doubling dose as the root of equation 8.6.6 is feasible in somatic cells, but nonlinearity of F(d) restricts its application.

A generation ago the linearity of radiosensitivity was supported by many experiments. However, at low dose rates chronic radiation has been found in Drosophila and the mouse to be only about one-fourth as effective as acute radiation. If the doubling dose for acute radiation is 40 rad, the doubling dose for chronic radiation is presumably about 160 rad (fig. 8.6.2).

Table 8.6.2: Induction of sex-linked lethals per acute roentgen

Species	Stage	K ($\times 10^{\circ}$)
Man	Oocytes	256
Rat	Spermatogonia	160
Mouse	Spermatogonia	186
Drosophila	Spermatozoa	24
Mammal		182

To evaluate the hazard of a mutagen, a constant population size is usually assumed. Then the average detrimental mutation leads to elimination through death or sterility of one zygote if selection is on heterozygotes and of half a zygote if selection is on homozygotes. Applied to one parent for one generation, a mutagen which induces U_R mutations per gamete that will be eliminated in homozygotes and U_D mutations per gamete that will be eliminated in heterozygotes leads ultimately to $U_R + 2U_D$ genetic deaths. This is also the incidence of genetic death if the mutagen is received by both parents for indefinitely many generations.

8.7 Approach to Equilibrium

Evolutionary genetics deals with the change of gene frequency per generation, which is usually of the form

$$\Delta = -k(q - Q) + 0(q - Q)^2, \tag{8.7.1}$$

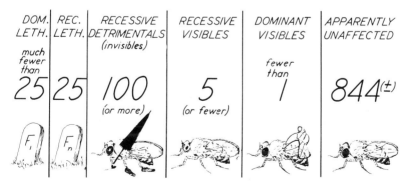

Fig. 8.6.2: Distribution of induced gene mutations among 1000 Drosophila spermatozoa irradiated with 150 r (after Muller).

the quadratic term being zero for mutation and simple models of migration, and negligible for selection in the neighborhood of the equilibrium gene frequency Q. Then k is called the (linear) *systematic pressure* or (coefficient of) recall (table 8.7.1). An alternative expression is

$$\Delta \doteq v(1 - q) - uq \qquad (8.7.2)$$

$$-\frac{\partial \Delta}{\partial q} = u + v = k,$$

which leads to the equilibrium $\Delta = 0$, or

$$Q = v/k \qquad (8.7.3)$$

At equilibrium the mean kinship for an array of demographically identical populations, not connected by migration, is to a close approximation

$$\phi \doteq \frac{1}{4Nk + 1} \qquad (8.7.4)$$

with gene frequency variance $Q(1 - Q)\phi$. The beta distribution of these gene frequencies was studied by Wright and others. It is characterized by two parameters that may be expressed in terms of Q and ϕ. Since they are estimable, the beta distribution is more useful in genetic epidemiology than the slightly more exact but complicated solution, which retains the quadratic term in equation 8.7.1.

When k is small we may interpret the change of gene frequency per generation as a derivative, $\Delta = \partial q/\partial t$. Rearrangement and integration gives

$$t \doteq \frac{1}{k}\left\{\ell n\left[\frac{q_0 - Q}{q_t - Q}\right]\right\} \qquad (8.7.5)$$

as the number of generations required under our simplifying assumptions to go from an initial gene frequency q_0 to a value q_t, where

$$q_0 < q_t < Q \quad \text{or} \quad Q < q_t < q_0.$$

The number of generations required to go halfway to equilibrium is therefore

$$t \doteq (\ell n 2)/k \qquad (8.7.6)$$

Some of the most interesting cases treat q_0 as an equilibrium fre-

Table 8.7.1: Changes in systematic pressure

Initial condition	Initial gene frequency (q_0)	New condition	Equilibrium gene frequency (Q)	Number of generations required to go halfway to equilibrium
I. Autosomal dominant disease				
Mutation-selection balance	u/s	$u \to cu$	cu/s	$(1/s)\ln 2$
Mutation-selection balance	u/s	$s \to s/c$	cu/s	$(c/s)\ln 2$
II. X-linked recessive disease				
Mutation-selection balance	$3u/s$	$u \to cu$	$3cu/s$	$(3/s)\ln 2$
Mutation-selection balance	$3u/s$	$s \to s/c$	$3cu/s$	$(3c/s)\ln 2$
III. Autosomal recessive disease				
Mutation-inbreeding balance	u/Fs	$F \to 0,\ h = 0$	$\sqrt{u/s}$	$(1/2Qs)\ln 2$
Mutation-inbreeding balance	u/Fs	$F \to 0,\ h > 0$	u/hs	$(1/hs)\ln 2$
Mutation-carrier selection balance	u/hs	$h \to 0$	$\sqrt{u/hs}$	$(1/2Qs)\ln 2$
Mutation-selection balance	$u/(h + q + F')s$	$F' \to F,\ F > F'$	u/Fs	$(1/Fs)\ln 2$
Mutation-selection balance	$u/(h + q + F)s$	$u \to cu,\ h + F \to 0$	$\sqrt{cu/s}$	$(1/2Qs)\ln 2$
Mutation-selection balance	$u(h + q + F)s$	$s \to s/c,\ h + F \to 0$	$\sqrt{cu/s}$	$(c/2Qs)\ln 2$
IV. Heterozygote advantage (v = selection coefficient against normal homozygote)				
Mutation-selection balance	$u/(h + v + F)s$	$h + F \to 0,\ v > q_0$	$v/(s + v)$	$[1/(2Qs + v)]\ln 2$
Selection-selection balance	$v/(s + v)$	$v \to 0,\ h + F \to 0$	$\sqrt{u/s}$	$[1/2Qs]\ln 2$
Selection-selection balance	$v/(s + v)$	$s \to s/c$	$cv/(s + cv)$	$[c/(2Qs + cv)]\ln 2$
Selection-selection balance	$v/(s + v)$	$v \to cv$	$cv/(s + cv)$	$[1/(2Qs + cv)]\ln 2$

u = mutation rate, s = selection coefficient against disease, h = dominance in carriers.

quency that is replaced by a new equilibrium at Q. Rare dominants and sex-linked genes respond quickly to a change in selection. Rare autosomal recessives evolve slowly and require thousands of years to go halfway to equilibrium when inbreeding, mutation rate, or homozygous selection is changed.

8.8 Systematic Pressure

Equation 8.7.6 shows that approach to equilibrium is governed by the systematic pressure, which must typically be small. The mean systematic pressure has been studied in contemporary populations and in evolutionary theory. Fisher showed that the opportunity for selection is measured by the Malthusian parameter

$$m_0 = \ell n F \qquad\qquad (8.8.1)$$

where F is the mean fitness of the population defined by equation 8.2.3. The mean systematic pressure per locus is

$$\bar{k} = \overline{\Sigma q_i k_i} = m_0/n \qquad\qquad (8.8.2)$$

where n is the number of loci, i is an allele with frequency q_i, and summation is over all alleles at a locus.

Various estimates of m_0 are in substantial agreement: (a) a few extremely fertile isolates have been reported, where the mean number of children born to women who survived to the end of the reproductive period was as much as 9.9, giving $m_0 < 1.60$. Under rigorous conditions many women die before or during the reproductive period, and many of their children do not survive to maturity. Large families are most susceptible to infant death, and a woman in a hunting and gathering society is unable to rear more than one young child at a time. Therefore the actual rate of increase is less than 1.60. (b) In recent generations the maximum rate of population growth in the United States was $m_0 = .62$ for a generation of 30 years. A slightly smaller value was observed on two Micronesian atolls following depopulation by a typhoon. Since accidental causes of death and infertility operate on these populations, the potential rate of increase is greater than .62. (c) Fisher used data from the British peerage to calculate that two stocks, differing by two standard deviations in genotypic fertility, would have a difference of 0.96 in their Malthusian parameters. (d) Crow developed an index of total selection defined by $I = N_{t+1}/N_t - 1$, where N_{t+1} is calculated on the assumption that every individual from a family of size s who survives the reproductive period has s children. Thus

$$m_0 < \ell n(I + 1) \tag{8.8.3}$$

The largest recorded value of I is 3.69 for the Peri of New Guinea, corresponding to $m_0 < 1.54$.

All of these methods lead to a value of m_0 in the neighborhood of unity for a human population under ideal conditions, with a value twice as great being inconceivable under the restrictions of a hunting and gathering society. By equation 8.8.2 the systematic pressure if there are 10,000 loci is $\bar{k} = 10^{-4}$.

In evolution the rate of amino acid substitution has been found to be about 10^{-9} per *covarion* (variable codon) per year for structural proteins, and perhaps six times as much for total DNA based on thermostability. For a protein with 500 amino acids and five generations per year, the substitution rate would be 10^{-7} per locus per generation. Since the rate of substitution equals the systematic pressure under mutation, and twice as much under selection, the mean systematic pressure in evolution may be about 10^{-7} per locus per generation.

We saw in 7.8 that kinship of racial groups bioassayed for polymorphisms is about 200 times as great as for morbidity and mortality in outcrosses. This suggests that systematic pressure on polymorphisms is much smaller, and may be more variable. The systematic pressure on a recessive lethal is

$$k = -\partial\Delta/\partial q \doteq u + m(q + h + F) = .01 \tag{8.8.4}$$

The systematic pressure on polymorphisms may be only about $k/200 = 5 \times 10^{-5}$, which is of the same order as the mutation rate.

Among the important properties of polymorphisms are specificity, intermittence, plasticity, and diminution. *Specificity* signifies that selection acts on a particular polymorphism at particular developmental stages and by particular agents, association of given alleles with which may be pronounced even when the systematic pressure is minimal. At other stages of the life cycle, selection may be reversed or negligible.

Intermittence implies that selection acts on a polymorphism strongly in some generations and imperceptibly in others. These may be called the active and passive phases of polymorphic selection. For organisms with a generation time of less than one year, seasonal selection imposes a regular periodicity on this intermittence. If intermittence is not cyclic, or if the cycle is long, the polymorphism may be nearly neutral for many generations. During this passive phase the possibility of determining selective forces acting at other times is remote, although disease associations not necessarily related to the principal selective pressure may remain significant.

Plasticity connotes the change in selective forces and genetic modifiers under different environments and gene frequencies. The greatest differentiation of racial groups bioassayed as kinship is for gene frequencies near one-half, which is consistent with different selective pressure in different environments. Fairly uniform selection against deleterious genes contrasts with variable selection on polymorphisms, which on the average over a large area or time may be nearly neutral.

The sensitivity of polymorphic gene frequencies to variations in selection leads to a paradox. On the one hand, selection may promote genetic modifiers, some closely linked to the system, which evolve into the "supergenes" of mimicry and histocompatibility. It seems likely that such a system, with perhaps different modifiers in different populations, will persist in the species as a permanent part of its genetic architecture. On the other hand, as a favorable mutation increases in frequency, the same selection for modifiers will reduce any harmful pleiotropic effects of the gene, initially in heterozygotes but also to some degree in homozygotes. As the gene frequency increases, this selection of modifiers will be more intense and the harmful pleiotropic effects in homozygotes less severe. If this modification is completely effective and the direction of selection is not reversed, the new gene will replace the previous wild-type alleles and establish a monomorphism, but if the selection for modifiers is less effective or the environment reverts toward the one under which the old allele was favored, the polymorphism may be maintained with a diminished systematic pressure. Degeneration into transient polymorphism (favoring either allele) and environmental changes or selection of modifiers that increase the fitness of homozygotes are two mechanisms leading to *diminution* of systematic pressure, both of them compatible with long persistent polymorphism.

These attributes of specificity, intermittence, plasticity, and diminution lead us to infer that many polymorphisms experience extremely small systematic pressure and are virtually noncontributory to the segregation load. This is especially probable in an organism like man with long, irregular environmental cycles and currently reduced mortality. Further support for small systematic pressure on polymorphisms comes from the inbred load, which in Drosophila is largely due to lethals. This is expected for a mutational load, but is incomprehensible for a segregation load.

Evidence for small systematic pressure has suggested to some evolutionary geneticists that many amino acid substitutions are neutral. Unfortunately the distinction between neutrality and weak selection founders on a logical disjunction: either independent populations of equal size and systematic pressure are at equilibrium for all loci under neutral mutation, or this stringent hypothesis is in some respects wrong. In nature, of course, the hypothesis cannot be satisfied: different populations and loci do not have identical parameters; if the species is not at equilibrium for a particular genetic system, then a region cannot be; and gene products differ both in the number of amino acids they contain and the precision with which changes can be recognized. The precise replication needed to make a specific test for neutral mutation is just not feasible.

Although neutrality has been a sterile hypothesis, there is a sense in which many amino acid substitutions are "quasi-neutral". Not only is the mean systematic pressure so small that estimation is out of the question, but the distribution of gene frequencies whose kinship is given by equation 8.7.4 becomes U-shaped when $k \ll 1/N$, with random fixation of different alleles. The world population of hominids 300,000 years ago has been estimated as at least 10^6. Thus systematic pressures less than 10^{-7} may be quasi-neutral in the species.

Neutral mutation gained support because of the false impression that rates of DNA substitution are uniform. On the contrary, evolution of DNA has been most rapid for sequences that are not transcribed. Perhaps systematic pressures on these sequences are no greater than the mutation rate, which by base-pair deletions and insertions can alter adjacent structural loci. Perhaps adaptive functions will be discovered for sequences that are not transcribed, but may regulate protein synthesis or protect structural loci from some kinds of mutation. It seems unwise to accept neutrality even for the weakest systematic pressures, since that hypothesis discourages search for adaptation and is in other ways unappealing: (a) Spontaneous mutation rates are far greater per unit of absolute time in lower organisms, being much more comparable per generation. This can be understood as selection for an optimum rate with increasing length of generation, but is unexplained under neutral mutation. (b) The rate of neutral mutation has never been directly estimated in any organism, nor has a single domonstrably neutral mutation been observed. Nearly all detected mutations are deleterious. (c) The systematic pressure on amino acid substitution is

$$k = U + ps \qquad\qquad (8.8.5)$$

where U is the mutation rate to neutral and advantageous genes and p is the proportion of advantageous substitutions with mean selection coefficient s. Given $s < 10^{-4}$ and U not demonstrably different from zero, there is no difficulty in explaining a nearly constant rate of amino acid substitution by a reasonable value of $ps > U$.

The principal role of evolutionary genetics in genetic epidemiology is to stress the smallness of systematic pressure and the slowness of evolution. It would be a serious error, at this stage of ignorance, to introduce consideration of long-term effects of relaxed selection into disease control.

8.9 Questions

1. *How are the affection risks a_i defined in section 4.8 for segregation analysis related to incidence and prevalence?*

If affection does not cause mortality, the a_i correspond to the specific risks m_k, which determine incidence, prevalence, and morbid risk. If affection causes mortality which is accurately ascertained in families of probands, then a_i is the probability of affection before the midpoint of i.

2. *How many individuals must be at risk to eliminate a gene with systematic pressure k?*

$1/k$

3. *Express indirect mutation rate estimates for single loci in terms of gene frequency.*

Autosomal dominant:	$u = mq$
Sex-linked recessive:	$u = xq$
Autosomal recessive:	$u = kq$

4. *Haldane's formula (equation 8.4.1) is disturbed by reproductive choice dependent on pregnancy outcome. Give examples of biases upward and downward.*

Termination of reproduction at birth of first affected child induces selection against carriers, and so x goes up. Continuation of reproduction to obtain a healthy male gives carriers greater fertility, and so x goes down.

5. *What are the three main constraints on these biases?*

(a) The $(n + 1)^{st}$ child is often conceived before the nth child is affected; (b) efficient contraception has been common only in the last generation; and (c) different couples pursue different reproductive strategies.

6. *Suggest a distribution to eliminate ascertainment bias toward familial probands.*

The conditional probability of carrier tests, given an isolated proband in the pedigree.

7. *Give an argument why the rate of amino acid substitution should be nearly constant if the systematic pressure leading to substitution is dominated by selection.*

Over long periods of time selection should be determined by a gradual change in the physical and biological environment, which is fairly constant.

8. *Show how the mutation rate is optimized by selection.*

Suppose that a proportion x of mutants have selective advantage s, and the remainder are detrimental. Since the probability of fixation of an advantageous mutant is nearly 2s, and a disadvantageous mutant in the process of elimination reduces fitness by $1/2N$, the net effect of a mutation rate u on fitness is $[x(s)(2s) - (1 - x)(1/2N)]u$. A higher mutation rate is favored if this quantity is positive, and a smaller mutation rate if it is negative.

9. *It has been suggested that the high mutation rate for Duchenne muscular dystrophy is due to recombination between two hypothetical loci, A and B, such that ab hemizygotes are affected and all heterozygotes type as carriers. Summarize evidence against this hypothesis.*

A. The *ab* haplotypes could arise only by recombination in *Ab/aB* females. Half of mutant and inherited haplotypes would be transmitted to infertile sons. The proportion of sporadic cases at equilibrium would be x = 1/2, much higher than observed.

B. All mothers of affected boys would type as carriers, whereas a significant excess are normal by comparison with obligate carriers.

C. Since *Ab* and *aB* hemizygotes are normal *ex hypothesi*, there is no reason why Lyonized *Ab/aB* women should be subclinically abnormal.

D. The excess of Duchenne muscular dystrophy among X-autosome translocations is consistent with a fragile site, but not with origin by recombination.

E. A large fraction of control women, being heterozygotes, should (but do not) type as carriers.

10. *Some geneticists emphasize the upward bias in estimates of drastic mutation rates and suppose that the mean is only 10^{-6}. If the induced rate is $1.7 \times 10^{-7}/r$, as for selected loci in the mouse, what would be the doubling dose?*

$1/.17 = 6$ r

11. *From table 8.6.2 and equation 8.6.4, estimate the number of lethal-bearing loci in man.*

For $K = 182 \times 10^{-6}$, $v = 2.6 \times 10^{-7}$, $g = .06$, we compute $n = 11,667$.

8.10 Bibliography

Barrai I, Mi MP, Morton NE, Yasuda N: Estimation of prevalence under incomplete selection. Am J Hum Genet 17: 221–236, 1965

Hook et al. (eds.): Birth Defects Symposium XI. Human Mutation: Biological and Population Aspects. Academic Press, New York, in press

McElheny VK, Abrahamson S (eds): Banbury Report 1. Assessing Chemical Mutagens: The Risk to Humans. Cold Spring Harbor Laboratory, New York, 1979

Morton NE, Lalouel JM: Genetic counseling in sex linkage. In: Birth Defects: Original Article Series, Vol 15(5C). Edited by Epstein CL. The National Foundation, New York, 1979, pp 9-24

National Research Council, Committee on the Biological Effects of Ionizing Radiations: The effects on populations of exposure to low levels of ionizing radiation. National Academy Press, Washington, DC, 1980

Templeton AR, Yokoyama S: Effect of reproductive compensation and the desire to have male offspring on the incidence of a sex-linked lethal disease. Am J Hum Genet 32: 575–581, 1980

Vogel F, Rathenberg R: Spontaneous mutation in man. In: Advances in Human Genetics. Vol 5. Edited by Harris H, Hirschhorn K. Plenum Publishing Corporation, New York, 1975

Winter RM: Estimation of male to female ratio of mutation rates from carrier-detection tests in X-linked disorders. Am J Hum Genet 32: 582–588, 1980

9. Cytogenetics

Genetic epidemiology of chromosomal abnormalities has been called *population cytogenetics*. It is concerned with frequencies, causation, and selective effects of chromosome aberrations, which are responsible for a surprisingly large fraction of mortality and morbidity in man.

9.1 Frequencies

Three populations are of special interest: live births, cases selected by disease or institutionalization, and spontaneous abortions. By combining data on the first and second groups, the relative risk of disease or institutionalization may be calculated as in table 6.3.1. By combining data on the first and third groups, the relative risk of spontaneous abortion and the incidence among conceptions may be estimated. Together with data on postnatal survival and reproduction, this permits calculation of fitness and, therefore, of mutation rates by the indirect method. Since misclassification of cytogenetically examined individuals is negligible, the direct estimate of mutation rates is also applicable when both parents of probands were examined (equation 8.3.1). If some parents were not examined, the semidirect method may be applied to the estimated proportion of sporadic cases (equation 8.3.4).

About 8 percent of recognized conceptions have a chromosomal abnormality that usually leads to spontaneous abortion. Perhaps chromosomal abnormality is even more frequent among conceptions that die in the first few weeks of gestation, only a small and biased fraction of which are successfully cultured for cytogenetic study (table 9.1.1).

9.2 Monosomy

Of all monosomies, only the 45,X type is found with appreciable frequency among recognized products of conception. They make up more than 9 percent of spontaneous abortions, but few are born alive. Post-

Table 9.1.1: Incidence of chromosomal abnormalities in three populations per 100,000

Constitution	Live births	Spontaneous abortions	Recognized conceptions[a]	Survival probability[b]
Normal male	51017	23474	46886	.925
Normal female	48380	26557	45107	.912
Trisomy	289	26299	4190	.059
Monosomy	5	9448	1421	.003
Polyploidy	2	11429	1716	.001
Structural	253	1981	512	.420
Other (mostly mosaic)	54	812	168	.273
Proportion abnormal	.006	.500	.080	

[a] Assuming that 15% of recognized conceptions are spontaneously aborted
[b] .85L/R, where L, R are live births and recognized conceptions per 100,000, respectively

natal survival is reduced. At maturity they manifest Turner's syndrome of short stature and ovarian dysgenesis. Since fertility is almost zero, all cases are mutants.

Spontaneous abortions that are 45,X are rarely mosaic. They arise by anaphase lag in the early zygote or by meiotic nondisjunction, but the distribution of paternal and maternal origin has not been determined. Reduced maternal age has been reported in some studies, but not others. In Drosophila and the mouse, monosomy can be induced by X-irradiation of spermatogonia, oocytes, and especially of early zygotes.

A large proportion of liveborn 45,X are mosaics. In nonmosaics 77 percent have a maternal X^M, indicating that the paternal X^P or Y chromosome was lost. Although nondisjunction during spermatogenesis cannot be excluded, the frequency of mosaicism suggests that preferential loss of the paternal sex chromosome may occur at an early cleavage division of the zygote.

In principle, homologous chromosomes may be distinguished by a *heteromorphism*, which is a heritable, cytogenetically recognized variant of a particular band revealed by any staining procedure. A common class of heteromorphism affects the size and pattern of centromeric hetero-chromatin (C-bands) when stained with Giemsa. Another common class of heteromorphism is defined when quinacrine makes the short arms of acrocentric chromosomes fluoresce, revealing variation in the size and brightness of stalks and satellites. Unfortunately, the X chromosome is rarely heteromorphic, and so the mode of origin of a particular 45,X individual usually cannot be determined cytogenetically. However, several X-linked polymorphisms are suitable, including the Xg^a blood group and the *G6PD* locus.

The excess of X^MO over X^PO is not as striking as it might appear. When a paternal sex chromosome is lost, whether X^P or Y, the egg will nearly always contain a maternal X and give rise to an X^MO zygote. About half the time when a maternal X is lost, the zygote will be YO and die in early gestation. Therefore, on the simplest assumption of an equal frequency of X^P and Y among successful sperm and equal loss of X^P and Y from the early zygote, the expected frequency of X^PO among 45,X zygotes is 1/3. The observed frequency of 23 percent X^PO among the small proportion of 45,X zygotes liveborn may implicate either preferential loss of the paternal X chromosome or a less frequent survival of X^PO zygotes to term.

The XO individual is a sterile male in Drosophila, where maleness is determined by a 2 : 1 ratio of autosomal genomes to X chromosomes. In mammals the XO is female, showing that maleness is determined by the Y chromosome. XO mice are fertile, unlike Turner syndrome in man.

Girls with Turner syndrome may have some intellectual impairment, especially in space-form perception, a factor of intelligence in which males tend to score higher than females. The cause of these differences, like all problems in behavior genetics, is poorly understood, but a sex-linked gene seems excluded by family studies as well as by the XY *vs* XO contrast.

Monosomy must contribute substantially to unrecognized pregnancy loss. If YO and X^PO fetuses have the same frequency, and if the ratio of X^PO : X^MO is the same among spontaneous abortions and livebirths, then for every 100,000 recognized conceptions there should be (.23) 1421 = 327 YO fetuses. Every mode of origin for trisomy should produce monosomy, and so for every 4190 cytogenetically detectable trisomies, 327 + 4190 = 4517 monosomies should be spontaneously aborted but not recovered for cytogenetic observation. Many other chromosomal abnormalities, single gene defects, and pathology due to maternal factors must also lead to clinically unrecognized spontaneous abortion, the total frequency of which must therefore be closer to 25 percent than to the 15 percent we have assumed for conceptions amenable to cytogenetic study. The proportion of conceptions that die because of cytogenetic abnormality may be as high as 1/8.

In somatic cells the variance in number of sex chromosomes increases, and the mean decreases with age. It is not known whether this measure of aging is predictive of aneuploid gametes, cancer, or other disease.

9.3 Trisomy

Trisomy may be classified by the extra autosome or sex chromosome. Trisomies may be single (46 + 1) or rarely double (46 + 1 + 1). Mosaicism occurs seldom among liveborn trisomies but approaches 10 percent of trisomic spontaneous abortions, probably as an artifact of cell culture. The partition in table 9.3.1 is therefore approximate.

Table 9.3.1: Incidence of trisomy

Constitution	Per 100,000 live births	Per 100,000 spontaneous abortions	Per 100,000 recognized conceptions	Survival probability
47, XYY ♂	48	25	45	.907
47, XXY ♂	48	321	89	.458
47, XXX ♀	51	120	61	.711
+A or B (1–5)	0	2083	312	0
+C (6–12)	0	4005	601	0
+D (13–15)	5	4686	707	.006
+16	0	7850	1177	0
+E′ or F (17–20)	12	2043	317	.032
+21	125	2483	479	.222
+22	0	2683	402	0
Total	289	26299	4190	

XYY trisomy is associated with increased height, reduced intelligence, and behavior problems in a proportion of cases. Survival in utero is the same as for normal males, but fertility may be reduced. Behavioral effects have been characterized by psychological tests as lack of emotional control and poor defense mechanisms against anxiety. Electroencephalogram (EEG) abnormalities have been reported. These studies were prompted by the observation that the frequency of XYY is increased among males in mental-penal institutions. The imprisoned minority are characterized by irresponsibility, not aggression, and their crimes are usually against property.

The XYY male may have no serious problems. Even under favorable circumstances, it would be difficult to identify the environmental or physiological conditions that promote normal behavior. Unfortunately such research has been a victim of vigilante genetics, which by agitation and harassment under the guise of protecting individual rights has curtailed studies that had been approved by peer and judicial review. Perhaps it will be some satisfaction to the XYY individuals who may suffer unnecessarily in the future, and to the society that shares the burden, that their rights have been overprotected.

Other trisomies do not appeal to vigilantes, since the presenting symptom is usually clinical rather than behavioral. XXY trisomy is called Klinefelter syndrome. Patients are males with atrophic testes and variably eunuchoid habitus. The fetal death rate may be elevated, mean IQ is reduced, and the risk of institutionalization for mental subnormality or crime is increased. Early therapy with male hormones leads to increased virilization and often to psychological improvement.

XXX females may have depressed IQ, increased risk of epileptic seizures, poor development of external sexual characteristics, and secondary amenorrhea. However, many are fertile and transmit a single X chromosome. An unbiased assessment of all sex chromosome anomalies must wait until longitudinal studies of cases ascertained at birth are complete.

Autosomal trisomies are sublethal. Occasionally trisomy 13 and 18 are liveborn with multiple malformations and severe mental retardation. Very rarely, trisomies 8, 9, or 22 survive to birth. Down syndrome, caused by trisomy 21, is characterized by mental retardation and multiple defects. It is by far the commonest trisomy at birth, and the only one compatible with survival beyond infancy. However, life expectancy is greatly reduced, and most affected fetuses are spontaneously aborted (fig. 9.3.1).

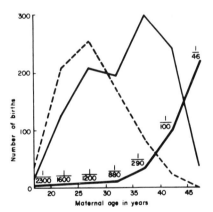

Fig. 9.3.1: The incidence of Down syndrome at birth. Key: --- all births in thousands; — mongols; — relative incidence.

The frequency of acrocentric trisomies increases dramatically with maternal age, both among spontaneous abortions and live births. A smaller effect of maternal age has been observed for most other trisomies.

Maternal exposure to X-rays has also been implicated, although the literature is not unanimous. Nondisjunction is increased by X-rays in somatic cells. X-irradiation of oocytes causes a low frequency of nondisjunction in the mouse, whose short generation time may give no valid comparison with the human female.

Table 9.3.2: Probabilities of the four modes of origin for trisomy

p = probability of paternal origin
  f = probability of first division nondisjunction
    given maternal origin
f + d = probability of first division nondisjunction
    given paternal origin

Origin	Probability
Paternal first division ($\mathcal{J}$I)	$p(f + d)$
Paternal second division ($\mathcal{J}$II)	$p(1 - f - d)$
Maternal first division ($\mathcal{Q}$I)	$(1 - p)f$
Maternal second division ($\mathcal{Q}$II)	$(1 - p)(1 - f)$

In recent years chromosomal heteromorphisms have revealed that most trisomies arise by nondisjunction at meiosis I in mothers. Less than 15 percent are paternal, mostly due to first division nondisjunction. Trisomies from young and old mothers appear to have the same origin (table 9.3.3). Heteromorphisms provide a powerful tool for genetic epidemiology, permitting separation of trisomies by parental origin (at least in probability) when testing specific maternal and paternal factors (table 9.3.4).

Table 9.3.3: Mating classes for a heteromorphic chromosome pair

Mating class	Description	Parent heterozygous	Father	Mother
A	2 chromosomes distinct, outcross	Neither	aa	bb
B	2 chromosomes distinct, intercross	Both	ab	ab
C	2 chromosomes distinct, backcross	Father	ab	aa
D	2 chromosomes distinct, backcross	Mother	aa	ab
E	3 chromosomes distinct, backcross	Father	ab	cc
F	3 chromosomes distinct, backcross	Mother	aa	bc
G	3 chromosomes distinct, intercross	Both	ab	ac
H	4 chromosomes distinct	Both	ab	cd

Table 9.3.4: Origin of trisomy

Mating class	Progeny type	Father	Mother	Child	♂I	♂II	♀I	♀II	Likelihood (L)
A	1	aa	bb	aab	1	1	0	0	p
	2			abb	0	0	1	1	$1-p$
B	3	ab	ab	aaa, bbb	0	1/2	0	1/2	$(1-f-pd)/2$
	4			aab, abb	1	1/2	1	1/2	$(1+f+pd)/2$
C	5	ab	aa	aaa	0	1/2	1/2	1/2	$(1-pf-pd)/2$
	6			aab	1	0	1/2	1/2	$(1-p+2pf+2pd)/2$
	7			abb	0	1/2	0	0	$p(1-f-d)/2$
D	8	aa	ab	aaa	1/2	1/2	0	1/2	$(1-f+pf)/2$
	9			aab	1/2	1/2	1	0	$(p+2f-2pf)/2$
	10			abb	0	0	0	1/2	$(1-p)(1-f)/2$
E	11	ab	cc	aac, bbc	0	1	0	0	$p(1-f-d)$
	12			acc, bcc	0	0	1	1	$1-p$
	13			abc	1	0	0	0	$p(f+d)$
F	14	aa	bc	abb, acc	0	0	0	1	$(1-p)(1-f)$
	15			aab, aac	1	1	0	0	p
	16			abc	0	0	1	0	$(1-p)f$
G	17	ab	ac	aab	1/2	0	0	1/4	$(1-p-f+3pf+2pd)/4$
	18			abb, bbc	0	1/2	0	0	$p(1-f-d)/2$
	19			abc	1/2	0	1/2	0	$(f+pd)/2$
	20			acc, bcc	0	0	0	1/2	$(1-p)(1-f)/2$
	21			aaa	0	1/4	0	1/4	$(1-f-pd)/4$
	22			aac	0	1/4	1/2	0	$(p+2f-3pf-pd)/4$
H	23	ab	cd	abc, abd	1	0	0	0	$p(f+d)$
	24			acd, bcd	0	0	1	0	$(1-p)f$
	25			aa–, bb–	0	1	0	0	$p(1-f-d)$
	26			–cc, –dd	0	0	0	1	$(1-p)(1-f)$

Each of the four columns under nondisjunction gives the conditional probability that, if nondisjunction occurs in the stipulated way and if the mating is of the given type, the child will be as indicated. Therefore the columnwise sums are unity for each mating class, as are the likelihoods.

9.4 Polyploidy

XXX and XXY triploids occur rarely among live births. Tetraploids and XYY triploids do not occur. All these polyploids are fairly common among spontaneous abortions. XYY triploids are least frequent, presumably because of early embryonic death. The frequencies of the sex chromosome types of triploids vary for no known reason among samples (table 9.4.1). The origin of nearly every polyploid can be determined from heteromorphisms. Most triploids are due to dispermy, although a substantial fraction are caused by suppression of the first meiotic division in either parent (table 9.4.2). Failure of either meiotic division in females is associated with reduced gestational age. The mechanism by which the mode of origin of the extra haploid set affects the time of abortion of a triploid conception remains obscure.

Table 9.4.1: Incidence of polyploidy

Constitution	Per 100,000 live births	Per 100,000 spontaneous abortions	Per 100,000 recognized conceptions	Survival probability
69, XXX	1	3419	514	.002
69, XXY	1	4933	740	.001
69, XYY	0	229	34	0
92, XXXX	0	1424	214	0
92, XXYY	0	1424	214	0
Total	2	11429	1716	

Table 9.4.2: Origin of triploidy

t = proportion from dispermy
p = probability of paternal origin of diploid gametes if not dispermy
f = probability of first meiotic division error if maternal diploid gamete
f + d = probability of first meiotic division error if paternal diploid gamete

Origin	Sex chromosomes	Autosomal heteromorphisms $\male$ = ab, $\female$ = cd	Marginal probability
2 Sperm, 1 egg (2$\male$)	XXX, XXY, XYY	(aa : 2ab : bb)(c : d)	t
Diploid sperm (1st division error), 1 egg ($\male$I)	XXY	(ab)(c : d)	$(1 - t)p(f + d)$
Diploid sperm (2nd division error), 1 egg ($\male$II)	XXX : YYY	(aa : bb)(c : d)	$(1 - t)p(1 - f - d)$
1 Sperm, diploid egg (1st division error) ($\female$I)	XXX : XXY	(a : b)(cd)	$(1 - t)(1 - p)f$
1 Sperm, diploid egg (2nd division error) ($\female$II)	XXX : XXY	(a : b)(cc : dd)	$(1 - t)(1 - p)(1 - f)$

Tetraploids are only of XXXX and XXYY constitution, suggesting that they arise by failure of an early cleavage in the zygote rather than by meiotic error.

No etiological factor for polyploidy has been established.

Table 9.4.3: Conditional probabilities for centrometric heteromorphisms in triploidy

Type	Father	Mother	Child	2δ	δI	δII	$♀$I	$♀$II
1	aa	bb	aab	1	1	1	0	0
2			abb	0	0	0	1	1
3	ab	ab	aaa, bbb	1/4	0	1/2	0	1/2
4			aab, abb	3/4	1	1/2	1	1/2
5	ab	aa	aaa	1/4	0	1/2	1/2	1/2
6			aab	1/2	1	0	1/2	1/2
7			abb	1/4	0	1/2	0	0
8	aa	ab	aaa	1/2	1/2	1/2	0	1/2
9			aab	1/2	1/2	1/2	1	0
10			abb	0	0	0	0	1/2
11	ab	cc	aac, bbc	1/2	0	1	0	0
12			acc, bcc	0	0	0	1	1
13			abc	1/2	1	0	0	0
14	aa	bc	abb, acc	0	0	0	0	1
15			aab, aac	1	1	1	0	0
16			abc	0	0	0	1	0
17	ab	ac	aab	1/4	1/2	0	0	1/4
18			abb, bbc	1/4	0	1/2	0	0
19			abc	1/4	1/2	0	1/2	0
20			acc, bcc	0	0	0	0	1/2
21			aaa	1/8	0	1/4	0	1/4
22			aac	1/8	0	1/4	1/2	0
23	ab	cd	abc, abd	1/2	1	0	0	0
24			acd, bcd	0	0	0	1	0
25			aa–, bb–	1/2	0	1	0	0
26			–cc, –dd	0	0	0	0	1

Table 9.4.4: Conditional probabilities for sex chromosomes in triploidy

Father	Mother	Child	Mechanism				
			2δ	δI	δII	$♀$I	$♀$II
XY	XX	XXX	1/4	0	1/2	1/2	1/2
		XXY	1/2	1	0	1/2	1/2
		XYY	1/4	0	1/2	0	0

9.5 Structural Abnormalities

Of all chromosomal anomalies, only those resulting from structural rearrangement within or between chromosomes are frequently inherited from carrier parents (table 9.5.1). Most structural abnormalities in live births are *balanced*: ie, euchromatic (nonheterochromatic) material is present in standard amount, without duplications or deficiencies. Most structural abnormalities in spontaneous abortions are *unbalanced*: some euchromatic regions are monosomic or trisomic.

Two acrocentric chromosomes can undergo Robertsonian translocation. If balanced the chromosome number is 45, the long arms are attached to one centromere, the heterochromatic short arms and one centromere are deleted, and structurally normal homologues are also present. If unbalanced, the chromosome number is usually 46 and a pair of one of the normal homologues is present, giving triplication for the long arm. Unbalanced Robertsonian translocations can arise either by mutation or by segregation in the carrier of a balanced translocation. A balanced homologous Robertsonian translocation, with two copies of an acrocentric long arm attached to the same centromere, arises by mutation but cannot be inherited in the balanced form, since all gametes are either nullisomic or disomic for that long arm.

Balanced reciprocal non-Robertsonian translocations can undergo adjacent or 3 : 1 segregation to yield aneuploid gametes. Inversion heterozygotes can give rise to aneuploidy by odd numbers of crossovers within the inversion. This is a rare event with some inversions, which may persist for many generations.

Unbalanced rearrangements include ring, marker, and supernumerary chromosomes. Breaks in different arms of the same chromosome can anneal into a *ring*. Besides deficiencies for two regions, such rings may undergo secondary breaks and losses in mitosis. Heterozygous deletions, detectable as a structurally abnormal *marker* chromosome with a normal homologue, are sometimes born alive, usually with multiple defects. Duplications may be present as a structurally abnormal *supernumerary* chromosome in addition to the normal diploid genome.

Data on reproductive fitness come from pedigrees ascertained through a balanced or unbalanced proband. Carriers of translocations have reduced fertility and an increased fetal death rate, reflecting segregation to unbalanced zygotes. The fitness of unbalanced translocations that cause spontaneous abortion is much less than pedigree studies on liveborn probands have indicated. Supernumerary chromosomes are eliminated largely through male infertility, not spontaneous abortion (table 9.5.2).

In a suggestive study most (11 of 13) non-Robertsonian structural rearrangements without additional chromosomes originated by mutation in a male. Most non-Robertsonian rearrangements with 47 chromosomes (7 of 7) originated by mutation in a female.

Table 9.5.1: Incidence of structural abnormalities

Constitution	Per 100,000 live births	Per 100,000 spontaneous abortions	Per 100,000 recognized conceptions	Survival probability (S)	Relative survival (S/.9185)[a]
Unbalanced					
Robertsonian translocation	5	876	136	.031	.034
Non-Robertsonian translocation	4	743	115	.030	.033
Ring, marker, and supernumerary	49	57	50	.833	.907
Balanced					
Robertsonian translocation	95	95	95	.850	.925
Non-Robertsonian translocation	84	172	97	.736	.801
Inversion	16	38	19	.716	.780
Total	253	1981	512		

[a] From table 9.1.1, .9185 is the mean survival of cytogenetically normal conceptions to birth, assuming 15% spontaneous abortion among recognized conceptions.

Table 9.5.2: Relative reproductive fitness and mutation rate of structural abnormalities

Group	Fitness from segregation and fertility	Fitness from survival to term	Incidence per 100,000 conceptions	Mutation rate/ gamete × 10^5
Robertsonian translocation	.788	.401[a]	231	69
Reciprocal translocation	.702	.384[a]	212	65
Inversion	.926	.780[a]	19	2
Ring, marker, and supernumerary	.323[a]	.907	50	17
Total				$\overline{153}$

[a] Used to estimate mutation rate

9.6 Chromosome Breakage

In table 9.5.2 the mutation rate for structural rearrangements is estimated as $V = 153 \times 10^{-5}$ per gamete per generation, based on all pregnancies that survive long enough to be clinically recognizable. Under stringent assumptions, this gives an estimate of the drastic mutation rate per locus. Suppose that all structural rearrangements survive to detection of pregnancy and are cytologically recognizable, and that half of multiple breaks lead to rearrangements, while others restitute. Then if there are n breakable segments, each containing one locus with breakage rate v, the Poisson distribution gives the expected frequency of multiple breaks as $1 - e^{-nv}(1 + nv)$, which for small values of nv approaches $(nv)^2$. On these assumptions

$$V = (nv)^2/2 \tag{9.6.1}$$

$$v = \sqrt{2V}/n.$$

Substituting $n = 10,000$, $V = 153 \times 10^{-5}$,

$v = 5.5 \times 10^{-6}$ breaks per locus per generation,

which is close to estimates of the drastic mutation rate. Evidently many (but not all) drastic mutants are due to breakage. A spontaneous mutation rate less than v is reasonably excluded.

Chromosome breakage can be a cause of neoplastic disease. Chronic granulocytic anemia is associated with the Philadelphia (Ph^1) chromosome, a translocation between chromosome 22 and (usually) chromosome 9. Such patients are chromosomally normal in all tissues except the hematopoietic system. Meningiomas frequently have a partial or complete deletion of chromosome 22. Patients with Down syndrome have an elevated risk ($\times 15$) for leukemia. Individuals with 13q-deletion are prone to retinoblastoma.

These and other examples are part of the evidence that neoplasia results from two or more genic or chromosomal mutations (spontaneous or induced by viruses, chemical mutagens, or ionizing radiation), which are selected in somatic cells because of an uninhibited growth rate. Any factor that promotes gene mutation or chromosome breakage is therefore likely to induce cancer, and vice versa.

The mutational theory of carcinogenesis is dramatically supported by a group of rare recessive diseases with hereditary chromosomal instability and susceptibility to neoplasia due to a defect in DNA repair. Fanconi anemia shows increased breakage with mitomycin C and alkylating agents, with reduced ability to excise DNA crosslinks. Heterozygous carriers are abnormally responsive to chromosome breakage by diepoxybutane. In Bloom syndrome the frequency of spontaneous sister chromatid exchange (SCE) is greatly increased. Homozygous cells have a retarded rate of DNA fork progression and are abnormally sensitive to UV-induced breakage. Cells from patients with ataxia telangiectasia exhibit a reduced ability to excise DNA bases damaged by ionizing radiation. Heterozygous cells are intermediate in their ability to survive X-irradiation. Several complementation groups of xeroderma pigmentosum hyperreact to induction of SCE by UV-irradiation and alkylating agents. A search for chromosome instability is being made in exceptional pedigrees with a high incidence of certain types of cancer.

An important class of chromosomal markers has been wrongly attributed to chromosomal fragility. By current techniques less than 1 percent of the population has *a fragile site* with the following features: a proportion of cells in appropriate culture conditions show a nonstaining gap of variable width, which usually involves both chromatids; the site is always exactly at the same point on the chromosome in cells examined from any individual patient or kindred; the site is inherited as a Mendelian dominant. Fragile sites appear to be a coiling defect, and breakage is atypical. It has been suggested that suitable techniques may reveal many fragile sites that differ in whether their frequency is enhanced by folic acid or thymidine deprivation, methotrexate, or elevated pH. Some viruses induce specific lesions that resemble fragile sites and are also useful markers for linkage. Only the site at Xq28 is known to be associated with disease, a form of mental retardation characterized by large testes and a distinctive facies and speech pattern. Carrier women with a high frequency of gaps may be mentally retarded.

9.7 Ovarian Teratoma

Benign teratomas of the human ovary are composed of a cyst filled with sebaceous material and a nodular growth that contains a wide variety of histological cell types, including hair, connective tissue, neural tissue, bone, teeth, and respiratory epithelium. All are of the 46, XX karyotype, and all chromosomes are maternally derived. Teratomas arise by abnormal development of a primary oocyte that is retained in the ovary. The second meiotic division is either suppressed, or polar body II fuses with the secondary oocyte. Centromeric markers heterozygous in the "mother" are invariably homozygous in the teratoma. Biochemical markers heterozygous in the "mother" may be either heterozygous or homozygous: retained heterozygosity is called *nonreduction*. The frequency y of nonreduction increases from zero at the centromere to a theoretical maximum of 2/3 for a distal marker, corresponding to a sample of two homologous genes without replacement,

$$y_{max} = \frac{\binom{2}{1}\binom{2}{1}}{\binom{4}{2}} = \frac{2}{3} \qquad (9.7.1)$$

A theoretical relation between the frequency of nonreduction and centromere map distance has not been derived, but data on Drosophila and Neurospora suggest an empirical function intermediate between

$$y = \tfrac{2}{3}[1 - (1 - 2\theta)^{3/2}] \qquad (9.7.2)$$

and

$$y = \tfrac{2}{3} \sin[\tfrac{3}{2} \sin^{-1} 2\theta]$$

where θ is the frequency of recombination between the marker and the centromere. Practical use of teratomas for centromere mapping is limited by problems of availability and contamination by ovarian cells.

9.8 Hydatidiform Mole

The placenta of an abnormal fetus may develop into a growth called a *hydatidiform mole*. The epithelium of the chorionic villi proliferates while the stroma of the villi undergoes cystic cavitation, resulting in a mass of cysts resembling a bunch of grapes. This has been recognized pathologically as an abnormal pregnancy, but only in recent years has the origin been revealed. All chromosomes of a mole are inherited from the father, and typically all markers are homozygous. Evidently a single sperm fertilizes an enucleate egg and doubles to give an isogenic 46, XX mole. Exceptionally, a holandric 46, XY mole occurs by dispermy or fertilization with a diploid sperm.

Moles are at risk to develop into choriocarcinoma, a malignant tumor that is fatal to the mother unless suitably treated. About half of choriocarcinomas arise in other ways: following abortion, during normal pregnancy, or from ectopic pregnancy or genital and extragenital teratomas. The factors that promote neoplasm are not understood, but homozygosity may be important. Triploid fetuses undergo trophoblastic proliferation which has been called a partial mole. It is not known whether such tissue is at increased risk to develop into choriocarcinoma.

9.9 Dominant Lethals

When gonia or gametes are exposed to ionizing radiation or other mutagens, a proportion of the F_1 die. Special techniques are required to analyze these dominant lethals, which are due to chromosome breakage. One elegant method uses wasps with haplo-diplo sex determination. Treated males mated to untreated females produce haploid sons whose chromosomes are from the mother and diploid daughters whose chromosomes come from both parents. Induced dominant lethals therefore act exclusively to diminish the number of females. In most other organisms, including mammals, the mortality of F_1 zygotes provides a measure of dominant lethal induction. The survival curve for ionizing radiation is approximately

$$S = e^{-(a + kD)} \qquad (9.9.1)$$

where D is the dose and k is the rate of dominant lethal induction per gamete per unit of radiation. This is a one-hit curve, suggesting that most dominant lethals are induced by a single electron tract. However, the possibility that selective mortality or repair of mutant cells distorts a two-hit curve into a single-hit one cannot be excluded.

Mortality of F_1 zygotes may be assayed in different ways, with correspondingly different dosage reponse. Litter size in the mouse, being variable and regulated by selection, does not accurately reflect deaths before and shortly after implantation. In the prenatal method females are dissected at a suitable stage of pregnancy and the numbers of corpora lutea as well as dead and living implanted embryos are counted. It is thus possible to estimate the proportion of prenatal deaths that occur before or after implantation.

Dominant lethals include both deletions and translocations. A proportion of translocations are viable and fertile in the balanced form, but cause early fetal death if unbalanced. If death occurs late enough so that it is not obscured by other factors determining litter size, it gives rise to semisterility. Obviously an assay of semisterile F_1 detects only a fraction of dominant lethals.

An even more indirect assay uses sex ratio from treated males. If dominant lethals more frequently involve an X than a Y chromosome, there should be a relative excess of sons in the F_1. However, deletion of part of the Y chromosome can be lethal, so this method measures only an unknown fraction of dominant lethals on the X chromosome.

Most dominant lethals act early in gestation. Therefore mutagenic effects expressed in the F_1 as stillbirths, juvenile death, and morbidity are a small proportion of induced mutation, most of which will be expressed gradually over many generations.

9.10 Sex Ratio

Strictly the ratio of males to females is *sex ratio*, while the frequency of males is the *sex proportion*. However, it is common to use sex ratio as a synonym of sex proportion. The secondary sex ratio refers to birth, and the primary sex ratio to conception.

In almost all human populations the proportion of males among births is greater than .5, but at a later age females predominate. Excess male mortality has many causes, and is not demonstrably related to sex-linked genes. Among chromosomally normal spontaneous abortions females are reported in slight excess, but this is at least partly due to contamination of cell cultures with maternal tissue. Among induced abortions and all recognized conceptions with normal chromosomes, there is an excess of males. Three explanations have been proposed: that females are more susceptible to early zygotic death; that Y-bearing sperm are formed in greater numbers than X-bearing sperm (meiotic drive) or are more successful in fertilization (gametic selection); and that XX zygotes are more prone to loss of the X chromosome than XY zygotes. XO and other aneuploids need not be considered, since they are excluded from chromosomally normal spontaneous abortions, and effectively from live births. Meiotic drive has been demonstrated for the SD locus in Drosophila and gametic selection for t alleles in the mouse. In man none of the explanations has been excluded, and so attempts to relate the sex ratio of abortions or live births to sex-linked genes have been unsuccessful. The possibility of meiotic drive or gametic selection is suggested by male-determined differences in sex ratio in the mouse, other mammals, and perhaps in man.

In at least some populations the proportion of sons declines slightly as paternal age increases, with maternal age and birth order standardized by covariance analysis. This is contrary to what would be expected if X-linked dominant lethals increase with the number of gonial divisions, but is consistent with gametic selection dependent on duration of the haplophase. Since the frequency of copulation declines with age, the intervals of storage in the male, survival in the uterus until ovulation, and postovulatory delay before fertilization must increase with age. If X-bearing sperm remain functional longer than Y-bearing sperm in any of these intervals, the proportion of males should decrease with paternal age, and conversely should increase with the frequency of copulation. This may explain the increase of sex ratio following certain wars.

The only clear prediction of a genetic effect on sex ratio is for induced recessive lethals, which should cause a decrease in sons born to women exposed to a mutagen (equation 8.6.3). Other sources of variation in sex ratio must be eliminated by appropriate controls, covariance analysis, and replication (as with all epidemiological studies). Unfortunately, little attention has been given in the mouse and man to this most sensitive and direct estimate of induced lethal mutation.

9.11 Acrocentric Associations

The short arms of acrocentric chromosomes in man consist of the basal short arm proper, a constricted region called the satellite stalk, and a terminal knob or satellite. In mitotic metaphases the acrocentric short arms tend to be associated (ie, to lie close together). This may be defined in various ways: one convention is to consider two short arms associated if the distance between their distal ends and the point at which their long axis intersects is less than the length of a G group chromosome. Conflicting claims have been made about frequencies and randomness of association, based partly on misunderstanding about the expected distribution of associations.

A theory of all possible combinations of acrocentric associations is not feasible. The problem can be made manageable by partitioning an association of n chromosomes into all $n(n - 1)/2$ possible pairs. The simplest theory assumes that all acrocentrics are equally likely to enter into association. Then if there are N acrocentrics (where $N = 10$ for a diploid, 11 for an acrocentric trisomy, and 8 for a balanced Robertsonian translocation), the probability under sampling without replacement that a pair of randomly associated chromosomes include i and j ($i, j = 1, \ldots N$) is

$$P_{ij} = \frac{1}{\binom{N}{2}} = \frac{2}{N(N - 1)}. \tag{9.11.1}$$

This hypothesis is readily excluded: on average, G group chromosomes are more likely to enter into association than D group chromosomes, and even homologues have different association possibilities. Abandoning equal association probabilities but keeping the assumption of randomness, the expected frequency, when the association probability for the ith set of homologues is p_i, becomes

$$P_{ij} = \begin{cases} \dfrac{\binom{P_i\,N}{2}}{\binom{N}{2}} = \dfrac{p_i(p_i - 1/N)}{1 - 1/N} & \text{for } i = j \\[3mm] \dfrac{\binom{P_i\,N}{1}\binom{P_j\,N}{1}}{\binom{N}{2}} = \dfrac{2p_i\,p_j}{1 - 1/N} & \text{for } i \neq j \end{cases}$$

(9.11.2)

This is a special case of Wright's general formula for associated pairs in terms of the p_i and a constant correlation F between associated pairs (equation 7.2.1), or

$$P_{ij} = \begin{cases} p_i^2 + p_i(1 - p_i)F & \text{for } i = j \\ 2p_i\,p_j(1 - F) & \text{for } i \neq j \end{cases}$$

(9.11.3)

Making this substitution in equation 9.11.2 we have

$$F = -1/(N - 1)$$

(9.11.4)

which gives $F = -1/9$ for five pairs of acrocentrics.

Next we relax the assumption that p_i is constant between homologues, but still assuming that F is constant and that association is random. Then by equation 7.2.5 for hierarchic correlation,

$$1 - F = (1 - F_W)(1 - F_A)\left(\dfrac{N}{N - 1}\right)$$

(9.11.5)

Here F_W is a negative fraction due to heterogeneity between homologues within an individual; F_A is a positive fraction due to heterogeneity of the p_i among individuals (equals zero if the p_i are estimated separately for each individual), and $-1/(N - 1)$ is the correlation between pairs drawn without replacement from a set of N. Likelihood ratio tests of hypotheses developed for factor-union phenotype systems in diploid individuals are directly applicable to pairs of associated chromosomes, bearing in mind that pairs from the partition of multiple associations are dependent. If heterogeneity in F between homologous pairs of chromosomes is significant, the generalized model which contains an F parameter for each pair of chromosomes is appropriate.

In a large sample F varied significantly among individuals, the mean being $-.124 \pm .007$. The deviation from $-1/9$ is in the expected direction, since F_W in equation 9.11.5 is negative. There is no clear evidence against random association. Apparently nonrandomness does not explain the excess of 13/14 Robertsonian translocations, which may be due to fragility of their short arms or to early fetal loss of other Robertsonian translocations.

The total frequencies of association vary among individuals, replicates, and observers. When these sources of variation are allowed for, there is no evidence that association frequency is related to risk for trisomy or morbidity, or to particular types of Q-heteromorphisms.

9.12 Questions

1. *From equation 9.6.1 develop a method to estimate w, the induced breakage rate/locus per generation, from the observed frequency V_D of structural rearrangements from parents with dose D of ionizing radiation,*

   $V_D = a + nwD/2$

2. *What is the doubling dose for chromosome breakage?*

   v/w

3. *Why is there an apparent deficiency of monozygotic twins with abnormal karyotypes?*

   The high death rate of fetuses with abnormal karyotypes makes it unlikely that affected MZ twins (concordant or not) will survive to recognition.

4. *Why must reports of impairment in sex chromosome anomalies be viewed with caution?*

   They are based on cases selected through disease, infertility, or behavior disorder, and are not representative of all cases.

5.   *What is the main use of heteromorphisms in research?*

To determine the origin of chromosomal anomalies. They are also of use in parentage tests and as markers for linkage.

6.   *What values of p_i in equation 9.11.2 correspond to equal association probabilities?*

$$p_i = \begin{cases} 2/10 \text{ for a diploid (N = 10)} \\ 3/11 \text{ for an acrocentric trisomy (N = 11)} \\ 2/8 \text{ for a balanced Robertsonian} \\ \quad \text{translocation (N = 8).} \end{cases}$$

7.   *Show that the above values of p_i satisfy equation 9.11.1 for pairs of homologues.*

For a diploid, $P_{ij} = 1/45$; for a balanced Robertsonian translocation, $P_{ij} = 1/28$.

8.   *Why do these values of p_i not satisfy equation 9.11.1 for trios of homologues and pairs of nonhomologues?*

In equation 9.11.1 each of the N chromosomes is assumed distinct. In equation 9.11.2 homologues are not distinguished.

9.   *Section 2.5 states that a proportion of tetraploid cells is often found in newly-established tissue cultures. Why does this not extend to long-established cultures?*

Clonal selection in culture favors aneuploidy.

10. *The vigilantes who curtailed longitudinal studies of sex chromosome abnormalities asserted the right of the people to determine scientific goals. Discuss.*

The people were a handful of Marxist activists and did not represent patients, their families, or providers of health care, who had most to lose from enforced ignorance.

9.13 Bibliography

De Grouchy J Turleau C: Clinical Atlas of Human Chromosomes. John Wiley, New York, 1977

Hamerton JL: Human Cytogenetics. Vol 2. Academic Press, New York, 1971

Hassold T, Matsuyama A: Origin of trisomies in human spontaneous abortions. Hum Genet 46: 285–294, 1979

Hook EB, Porter IH (eds): Population Cytogenetics. Studies in Humans. Academic Press, New York, 1977

Jacobs PA, Angell RR, Buchanan IM, Hassold TJ, Matsuyama AM, Manuel B: The origin of human triploids. Ann Hum Genet 42: 49–57, 1978

Kajii T, Ohama K: Androgenetic origin of hydatidiform mole. Nature 268: 633–634, 1977

Linder L, McCaw BK Hecht F: Parthenogenic origin of benign ovarian teratomas. N Engl J. Med 292: 63–66, 1975

Sutherland GR: Heritable fragile sites on human chromosomes. Am J Hum Genet 31: 125–148, 1979

10. Genetic Risks

Genetic epidemiology, like other branches of epidemiology, has concentrated on the etiology and distribution of disease as a foundation for possible control measures. The principal application has been to genetic counseling.

10.1 Genetic Counseling

Genetic counseling can be resolved into three distinct operations. In the first, the problem is stated by *diagnosis* of affection in family members. Usually the diagnosis must be extremely precise: for example, there are many simply inherited forms of cleft palate, which for genetic counseling should be distinguished from the residual class of obscure etiology. The second operation is *prognosis* for the disease in patients and for recurrence risks in relatives. The final step is *communication* of medical and genetic information to the patient or his relatives. Subsequently medical, social, or psychological services may be offered.

Clinical geneticists naturally concentrate on diagnosis, and paramedical genetic counselors on communication. However, the precise prognosis of genetic risks is the central problem of genetic counseling, without which the transaction is mere charlatanism. It is remarkable that no standards have been set, and even the theory is poorly developed.

Recently members of the European Society of Human Genetics were presented with some problems on recurrence risks for sex-linked lethals. Most of the members of the society did not attempt the test: of those who did, the majority made gross errors. This was a particularly easy situation, since assumptions were clearly stated and there were no carrier tests. In practice, uncertainty about fitness, mutation, and the distribution of carrier tests raises more complicated problems. Beyond these straightforward Mendelian cases, in the area of complex inheritance the specification of risk is still more difficult. Medical geneticists are now being licensed, but this does not provide quality control of the risk prevision that operationally defines genetic counseling.

To clarify the issues some definitions are required: *genetic counseling* is a transaction in which information I about the genotype of an individual is used to provide an estimate of disease risk, called a *genetic risk* in general, but a *recurrence risk* if the information includes at least one affected relative. The individual at risk is called the *root* of his pedigree. Genetic risks are of three kinds, denoted $R(g_i)$, $R(p_i)$, and $R(p_{ij})$. The probability that the root be of genotype g_i is

$$R(g_i) \equiv P(g_i \,|\, I)$$

This is defined for a major locus genotype, not a polygenic one. The root may be included in I. The probability that the root be of phenotype p_i is

$$R(p_i) \equiv P(p_i \,|\, I)$$

In this case the root is not included in I. The probability that a root which can take liability indicator j be of phenotype p_i is

$$R(p_{ij}) \equiv P(p_{ij}) \,|\, I)$$

This is the most complex risk, since under certain conditions the root at a different liability indicator $h \neq j$ may be included in I. For example, we may require the probability that a member of a pedigree for a dominant disease of late onset, like Huntington's chorea, who is known to be unaffected at age h, be affected before age $j > h$. The general condition for equation 10.1.3 to be defined is that a transition from p_{ih} to p_{ij} be possible, which implies: (a) if p_{ij} is normal, p_{ih} must be normal; (b) if the liability indicator is temporal, j must be later than h.

In the best case the genotype of the root can be established with certainty by chemical, biochemical, or cytogenetic criteria and the genetic risk can be reduced. For example, individuals of genotype U/U at the $E1$ pseudocholinesterase locus are acutely sensitive to the anesthetic suxamethonium, to which they should not be exposed. Hemolytic anemia will occur if individuals with G6PD deficiency ingest a quinone, such as the antimalarial drug primaquine. Individuals of O blood group should never be transfused with red blood cells of type A, B, or AB. Pharmacogenetics looks to the day when every individual is characterized by idiosyncratic sensitivity to environmental agents he should avoid.

The situation is less satisfying if the genotype of the root can be established with certainty, but genetic risk cannot be reduced. If the root is a fetus, the parents may elect induced abortion for a sufficiently grave disease, or prompt treatment otherwise. If the root is liveborn, those responsible for his treatment may withhold intensive care if the condition is hopeless. These options raise medical, ethical, and legal problems, which are no different than for a disease of purely environmental etiology.

In many cases the genotype of the root cannot be established with certainty. Often the etiology is mixed, and the genotypes leading to affection cannot be completely resolved. A characteristic problem of genetic counseling is to predict disease risk in a child not yet conceived, as a guide to reproductive choice. Contraception, artificial insemination, elective abortion, and prompt treatment are alternative strategies, depending on risk, severity, and parental attitudes.

Although the main stream of genetic counseling estimates a risk R which is communicated to the attending physician, a minority view is that the risk conveyed to the root should be deliberately "fuzzy" (ie, categorized as high or low). Stevenson, Davison, and Oakes address this heresy:

> "It has been argued by many that it is not necessary to give a risk figure, but terms like 'high,' 'not high,' or 'low' are sufficient. It is difficult for the writers to conceive that the patients they see would accept such general and relative terms. ...In our opinion the single figure is not only the best way of expressing the true situation, but it is most easily understood by those who are asking the question."

Another view is that genetic counseling should avoid a model of inheritance unless it has been shown to be the only possible explanation for the disease in question, and instead use empirical risks that assume no model. Empirical risks depend on no detailed analysis, consider only the child immediately following a proband, and pool families of different compositions, ignoring normal siblings, more remote relatives, sex, age, quantitative information, and etiological heterogeneity. Instead of a specific risk R for unique information I, based on analysis of all available information, the empirical risk is based on an arbitrary subset of the data. Murphy and Chase discuss the deficiencies of empirical risks with respect to efficiency, homogeneity, precision, sufficiency, and universality, recalling the dictionary definition of "empiric" as "one who deviates from the rules of science or accepted practice; one who relies upon practical experience alone, disregarding all theoretical and philosophical considerations; hence a quack." An analytical risk (called modular by Murphy and Chase because it is based on a model of inheritance) is superior in all these respects but requires that the assumed mechanisms and parameters of inheritance be reasonably close to the truth. "Reasonably close" means that no other model satisfying the criteria of support, economy, and resolution (section 4.2) gives a substantially higher likelihood in a large body of carefully collected family material. The outcome of any segregation analysis implies a specific genetic risk for every conceivable pedigree, determined with the same precision as the segregation parameters. Obviously genetic risks cannot be more reliable than the segregation analysis on which they are based, and so careful attention must be paid to detecting weakness and introducing improvements in segregation analysis. The mixed model is biologically meaningful and so provides a general method. The parameters of the generalized transmission model (section 5.9) are biologically meaningless except in the mendelian special case, and so are not suitable for genetic risks.

Unfortunately, it will be a long time before adequate segregation analysis is performed for many diseases. Meanwhile, incomplete data on incidence and recurrence risks must be used as well as possible. Risks based on pairs of relatives can be fitted by a major locus, polygenic, or cultural model, usually without resolving these and more complex possibilities. In practice the single locus model will be used where it is supported, and the polygenic model otherwise. Communication of genetic risks must stress the ambiguity of the evidence and should give more than one risk when alternative models are tenable.

10.2 Computer Assistance

The three basic risks for genetic counseling are posterior (or inverse) probabilities that can be evaluated by Bayes' theorem. For example, equation 10.1.2 uses the identity

$$P(p_i \mid I) \equiv \frac{P(p_i, I)}{P(I)} \tag{10.2.1}$$

Lalouel and earlier authors have considered algorithms for this calculation. Excluding the root, a pedigree may be decomposed into founders (F) who have no ancestors in the pedigree and cognates (C) who have one or more ancestors in the pedigree. Let J be the set of liability indicators for the pedigree. Then

$$P(I) = \sum_F P(g_F) P(p_F, p_C \mid g_F, J) \tag{10.2.2}$$

$$P(p_i, I) = \sum_F P(g_F) P(p_F, p_C, p_i \mid g_F, J)$$

Equations 10.1.1 is evaluated in the same way, with g_i substituted for p_i.

To evaluate equation 10.1.3, let a_{ij} designate an affected root with liability indicator j. Then

$$R(p_{ij}) = \frac{P(a_{ij},I) - P(a_{ih},I)}{P(I) - P(a_{ih},I)}$$

(10.2.3)

where

$$P(a_{ij},I) = \sum_F P(g_F)P(p_F, p_C, a_{ij} \mid g_F, J)$$

It is feasible to compute genetic risks by hand for major loci with complete penetrance. Murphy and Chase give many examples. For instance, the genetic risk for a rare, completely penetrant recessive disease due to a single locus is P for children of an affected person, P/3 for children of normal sibs, P/2 for children of carriers, and P/4 for children of sibs of carriers, where

$$P \doteq 2F + q + h$$

(10.2.4)

and F is the inbreeding coefficient, q the gene frequency, and h the penetrance in heterozygotes. The risk is small except with inbreeding. Such calculations are simplified by numerical approximation and by ignoring information in the pedigree.

As the pedigree and/or mode of inheritance becomes more complex, precise calculation falls outside the competence of most genetic counselors, and in the limit would deter a mathematician. Computational errors are likely for complex pedigrees, even under simple modes of inheritance. Either genetic counseling must accept a low level of reliability for its risks, or demand a high level of mathematical sophistication in its counselors, or resort to computer assistance. The last alternative seems preferable. The same parameters and assumptions that give the best segregation analysis can be used to evaluate risks. As more diseases are submitted to complex segregation analysis, their parameters can be catalogued in the library of the computer program that calculates risks. All that the genetic counselor need do is to verify that the parameters are applicable to his case, enter the data, and review the calculated risk. These operations are conveniently done at a terminal linked by telephone to a computer that will receive the information interactively and return a computed risk in seconds. Such computer-assisted genetic counseling was made feasible by the recursive formulation of Murphy and Mutalik, implemented by Heuch and Li in a computer program, and extended to the mixed model by Lalouel. Integration with a data base was pioneered by Williams, and is still in progress.

10.3 Problems in Determining Genetic Risks

Computer assistance cannot solve all problems in determining genetic risks, which include specificity of the diagnosis, reliability of information, etiological heterogeneity, and obscure modes of inheritance.

The specific diagnosis should be known with certainty. If it is not, we may estimate the overall (or marginal) genetic risk as

$$R(p_i) = \sum_k c_k R_k(p_i) \qquad\qquad (10.3.1)$$

where $R_k(p_i)$ is the risk for the kth diagnosis and c_k is the probability that the kth diagnosis be correct, which may be analyzed further into

$$c_k = \frac{f_k P_k(I)}{\sum_k f_k P_k(I)}, \qquad\qquad (10.3.2)$$

where f_k is the prior probability of the kth diagnosis and $P_k(I)$ is the conditional probability of the information. If the f_k cannot be estimated objectively (and they rarely can), the R_k should be detailed with an attempt to identify the most likely, next most likely, and so on.

If the affection status of a relative is uncertain, it is helpful to compute risks given that the relative is affected or of unknown phenotype. However there can be no adequate correction for unreliable data, and every risk is conditional on the validity of the information.

Etiological heterogeneity is similar to an uncertain diagnosis. Equation 10.3.1 may be used, with R_k the risk for the kth mode of inheritance and f_k the probability of that mode, among roots with the same diagnosis coming to counseling. Thus f_k may be different from the population frequencies, and so difficult to determine. It is usually necessary to assume that f_k is the same for roots, probands, and cases in the general population. The probability that a pedigree belong to the kth mode of inheritance is not in general equal to f_k, which is defined on affected individuals. If segregation analysis revealed different parameters among mating types due to etiological heterogeneity, parameters specific for parents of the root should be used, and the f_k should refer to cases from that mating type. Usually segregation analysis tries to discriminate dominant and sex-linked pedigrees, leaving recessive, polygenic, and other cases of obscure etiology as a residual under the mixed model. The same partition should be made for genetic risks by equation 10.3.1.

Confidence is warranted in genetic risks based on segregation analysis of a large, sound body of data in which goodness of fit tests are satisfied. As the quantity and quality of segregation data decline, different modes of inheritance give an adequate fit, and the etiology becomes obscure. There is little alternative to presenting risks for the best-fitting simple models. The polygenic model with one or two parameters usually qualifies, as well as some major locus model with d fixed at 0 or 1 and two free parameters (t and q). Because the parameters are estimated from segregation data, they give similar risks for most roots. When risks from two acceptable models are widely discrepant, interpretation must be as cautious as when the diagnosis is in doubt.

Although different models often give comparable risks when the root is a singleton raised by his own parents, unusual situations like adoptions, half sibs, and monozygotic twins raise special problems when etiology is so obscure that maternal effects and cultural inheritance have not been excluded. The genetic counselor must be especially cautious in such cases, and not rely blindly on any risk, whether empirical, computer-assisted, or estimated by a trusted mathematician. The attraction of computer assistance is not that it is infallible, but that no other method can be consistently better.

10.4 Mutagens

Evaluation of genetic risks is not limited to counseling, but includes public health aspects of induced mutation. Table 10.4.1 estimates for a doubling dose given to both parents in a single generation the hazard in their progeny and descendants, assuming constant population size. Even a multiple of the doubling dose is a relatively small hazard to the first generation offspring, but the total number of genetic deaths is considerable. Most of them occur in generations far removed from the induced mutation, and the manner of genetic death (which includes not only overt morbidity but also unrecognized fetal loss, induced abortion, selective contraception, and involuntary sterility) is obscure.

Mutagenic hazards have a peculiar legal status. In the United States the Delaney clause prohibits food additives that have been shown to be carcinogenic at any dose in any mammal. Discretion is given to the Surgeon General to restrict drugs that are carcinogenic. While mutational hazards are not explicitly considered, nearly all mutagens are in some degree carcinogenic and *vice versa*.

Public health aspects of induced mutation are sometimes expressed as a "maximum permissible dose," a concept that poses the same logical and ethical problems as establishing a maximum permissible frequency of rape: what seems permissible to an expert committee may be received

Table 10.4.1 : An estimation of the hazards from a doubling dose of mutagen to a single generation

$(u = 21 \times 10^{-6}$ per locus per generation)

	Number of loci	Proportion of mutations expressed in F_1	Mutation rate per 10^6 zygotes	Genetic load (after Slatis)		
				Unrecognized fetal death (38%)	Observed fetal death (6%)	Postnatal death or abnormality (56%)
Autosomal dominant	1,000	.5	42,000	7,980	1,260	11,760
Autosomal recessive	10,000	.127 (.16)	420,000	3,243	512	4,779
Sex-linked recessive	600	.333 (.16)	18,900	383	60	564
Trisomy	—	1	4,190	?	3,944	246
Monosomy	—	1	5,938	4,517	1,417	4
Polyploidy	—	1	1,716	?	1,714	2
Structural	—	.61	3,060	?	1,023	170
			495,804	16,123	9,930	17,525
Total in F_1 assuming recessives expressed as infertility			—	12,880	9,418	12,746
Total in F_2			—	7,873	1,661	11,670

with much less favor by potential victims. The notion of balance between good and ill effects is not readily applied when the good effects are measured in terms of immediate gratification and the bad effects are experienced by future generations. This is a deeper problem than the long-term effects of genetic counseling, which in the worst possible case of incurable disease merely influences the time and manner in which an inevitable genetic death occurs. Genetic epidemiology should determine the consequences of induced mutation as accurately as possible, but should not attempt to fix maximum permissible doses.

10.5 Questions

1. *An older convention determines genetic risks for the hypothetical child of a marital pair of "consultands." In what way is this unsatisfactory?*

   Equation 10.1.3 for the risk to an individual observed at some stage to be normal is undefined.

2. *Would the presentation of risks for two or more tenable models be disturbing?*

   Limited experience supports the expectation that two hypotheses that fit a substantial body of data necessarily predict closely similar risks for all the commonly observed segregation patterns. Widely discrepant risks for acceptable models faithfully reflect the uncertainty of counseling in exceptional pedigrees.

3. *Combine equations 10.3.1 and 10.3.2.*

   $$R(p_i) = \sum_k f_k P_k(p_i, I) / \sum_k f_k P_k(I)$$

4. *Enormous effort has been devoted to mutation research, yet we know little about hazards to man. Why?*

   Research has been largely directed to the molecular basis of mutations, not to their epidemiological effects.

10.6 Bibliography

Epstein CJ, Curry CJR, Packman S, Sherman S, Hall BD (eds): Risk, communication and decision making in genetic counseling. Birth Defects: Original Article Series. Vol 15(5C). AR Liss, New York, 1979

Fuhrmann W, Vogel F: Genetic Counseling. Second edition. Springer-Verlag, New York, 1976

Lalouel JM: Probability calculations in pedigrees under complex modes of inheritance. Hum Hered 30: 320–323, 1980

Lubs HA, De La Cruz F: Genetic Counseling. Raven Press, New York, 1977

Murphy EA, Chase GA: Principles of Genetic Counseling. Year Book Medical Publishers, Chicago, 1975

Stevenson AC, Davison BA, Oakes MA: Genetic Counseling. JB Lippincott, Philadelphia, 1970

Williams WR: Computer-assisted genetic counseling. PhD dissertation, University of Hawaii, 1981

11. Surveillance

Surveillance (also called monitoring) implies recurrent measurement of any continuous process in order to take some course of action when a change appears to have occurred. This action necessarily includes supplementary tests to verify the reality of the change and then to establish its cause, as a basis for measures to eliminate or reduce the hazard. Surveillance has much in common with industrial quality control, which uses the sequential analysis that has proved so useful for linkage detection. Different approaches are appropriate for group comparisons, temporal increase, and clustering in time and space, with adaptation to random samples and to registers of affected. The biological hazards to be detected include mutagens, teratogens, and carcinogens.

Except for somatic mutations, surveillance for a mutagenic hazard requires a sentinel phenotype of low fitness and a correspondingly high proportion of new mutants. This condition is satisfied by all aneuploids and by certain autosomal dominants and sex-linked genes. Most idiomorphs are unsuitable, since only a negligible fraction are new mutants. For example, suppose that the mutation rate to a private allele is 10^{-6} per locus per generation, the doubling dose is 40 r, the rate of false paternity is .01, and the efficiency of parentage exclusion by genetic evidence is .99. Then in a population in which one parent was exposed to 200 r the probability that a sporadic case, not excluded on genetic evidence, be due to mutation would be about

$$\frac{(7 \times 10^{-6})}{7 \times 10^{-6} + .01\,(.01)} = 0.07$$

This is probably an overestimate, since mutagenic hazards are likely to be smaller than 200 r, parentage errors greater, and the efficiency of parentage exclusion less. Parentage errors are *a priori* unknown and likely to vary among groups within a population, and therefore between exposed and control parents. A surveillance program based on polymorphisms is necessarily prospective, and therefore costly. On the contrary, surveillance for aneuploidy and deleterious mutations can be either retrospective or prospective, and in favorable cases can exploit induced and spontaneous abortions in which the yield of mutants may be higher than for live births.

11.1 Environmental Hazards

Suppose a specific environmental agent has been suggested as a biological hazard. The problem is to test this hypothesis and, if confirmed, to establish the response to a given dose.

Let $P_d(X_i)$ be the probability that the ith individual or sample receiving a dose d take a random variable X_i. Then

$$z_i = \log [P_d(X_i)/P_0(X_i)] \tag{11.1.1}$$

is a lod score testing the hypothesis of no biological hazard and

$$\log B < \sum_i z_i < \log A \tag{11.1.2}$$

defines a sequential test with type I error

$$P(\Sigma z_i > \log A \,|\, H_0) = \alpha \tag{11.1.3}$$

and type II error

$$P(\Sigma z_i < \log B \,|\, H_1) = \beta \tag{11.1.4}$$

where

$$A \doteq \frac{1-\beta}{\alpha}, \ B \doteq \frac{\beta}{1-\alpha}. \tag{11.1.5}$$

It seems reasonable to take $\alpha = \beta = .1$, and therefore $A = 9$, $B = 1/9$. Then a "significant" response justifies suspicion but not certainty. As A increases we have more confidence that the response is real, providing $P_0(X_i)$ has been correctly specified.

Consider first a sentinel phenotype which may be caused by the hazard. Then if the samples are random we have the binomial distribution, with $X_i = 1$ if affected and 0 if normal. A linear dose response would give

$$P_d(X_i) = \begin{cases} (1 + Kd)p & \text{if } X_i = 1 \\ 1 - (1 + Kd)p & \text{if } X_i = 0 \end{cases}. \tag{11.1.6}$$

Radiation gives log linearity,

$$P_d(X_i) = \begin{cases} 1 - e^{-Kd}(1-p) & \text{if } X_i = 1 \\ e^{-Kd}(1-p) & \text{if } X_i = 0 \end{cases}. \tag{11.1.7}$$

If the outcome $X_i = 1$ is sensitive to the hazard and $X_i = 0$ is resistant, as for example if the former denotes the male and the hazard is maternal irradiation,

$$P_d(X_i) = \begin{cases} \dfrac{pe^{-Kd}}{1 - p + pe^{-Kd}} & \text{if } X_i = 1 \\ \dfrac{1-p}{1 - p + pe^{-Kd}} & \text{if } X_i = 0 \end{cases}. \tag{11.1.8}$$

The number of possible responses is indefinitely large and $P_d(X_i)$ must be determined for each biological hazard.

When surveillance is for affected individuals in a register that excludes normals, the distribution of X_i affected may be Poisson,

$$P_d(X_i) = \frac{\{(1 + Kd)m\}^{X_i} e^{-(1+Kd)m}}{X_i!}, \tag{11.1.9}$$

where m is the expected number of affected when $d = 0$.

If the event under surveillance is not an affected individual, but rather a cellular event like sister chromatid exchange which may be observed many times in a sample of N cells from an individual, the variation around the mean must be allowed for. Let σ^2 be the variance among individuals with the same dose. Then the probability of X_i events in a sample of size N may be normal,

$$P_d(X_i) = \frac{1}{\sqrt{2\pi}\,\sigma}\, e^{-(X_i - (1+Kd)m)^2/2\sigma^2}. \tag{11.1.10}$$

More generally, terms like $(1 + Kd)$ may be replaced by $f(d)$, a specific response.

11.2 Increasing Risks

Often surveillance is directed against unspecified hazards that may increase with time. The appropriate distribution for a sentinel phenotype is

$$P(X_i) = \begin{cases} Kp & \text{if } X_i = 1 \\ 1 - Kp & \text{if } K_i = 0 \end{cases}. \tag{11.2.1}$$

If the outcome $X_i = 1$ is sensitive to the hazard and $X_i = 0$ is resistant,

$$P(X_i) = \begin{cases} \dfrac{pK}{pK + 1 - p} & \text{if } X_i = 1 \\[2ex] \dfrac{1 - p}{pK + 1 - p} & \text{if } X_i = 0 \end{cases} \tag{11.2.2}$$

If surveillance is of affected individuals only,

$$P(X_i) = \frac{(Km)^{X_i} e^{-Km}}{X_i!}. \tag{11.2.3}$$

This is illustrated for Down syndrome in table 11.2.1. Finally, if X_i varies normally among individuals with mean Km and variance σ^2,

$$P(X_i) = \frac{1}{\sqrt{2\pi}\sigma} e^{-(X_i - Km)^2/2\sigma^2}. \tag{11.2.4}$$

Table 11.2.1: A sequential test on Down syndrome in Sweden ($\alpha = \beta = 0.1$, $K = 1.116$)

Year	n_i	m_i	Number of liveborn children	Z	Decision
1968	165	144.3	113,087	1.37	Continue
1969	131	137.4	107,662	−0.19	Continue
1970	142	140.6	110,150	−0.92	Continue
1971	135	144.9	113,512	−2.91	Accept H_0
1972	137	142.0	111,253	−1.44	Continue
1973	138	138.8	108,749	−2.39	Accept H_0
Total	848	848	664,413		

11.3 Clustering in Time

The distributions considered in the previous section are appropriate for a hazard whose duration may be long relative to a period of observation. However, some hazards lead to a cluster of cases in a relatively short time interval, with no seasonal or other regular recurrence. If the sentinel phenotype is rare, the cluster may be of such small size that usual tests of significance are unreliable. Often the main objective is to detect an increase in a sentinel phenotype, and then tests for clustering are supplementary.

Suppose first that the data file is an ordered register containing both affected and normals. For example, the register could consist of kary-otyped spontaneous abortions with affection defined as a chromosomal abnormality, or it might be a file of live births and registered fetal deaths. In both cases, the appropriate order would be by estimated date of conception.

On the hypothesis H_0 of no clustering, the probability that the next abnormal observation occur n events after the last abnormal is a geometric distribution,

$$P_0(n) = p(1 - p)^{n-1} \qquad (0 < p < 1, n > 0) \tag{11.3.1}$$

with mean waiting time

$$E_0(n) = 1/p. \tag{11.3.2}$$

On the hypothesis H_1 of clustering, we suppose two distributions, with risks p/K and pK, respectively. For $K > 1$, the first distribution corresponds to no epidemic and the second to an epidemic. The sequence of trials leading to an abnormal is taken at random from one of the two distributions, with probability $1/(K + 1)$ for the first. Since the waiting times are K/p and 1/pK, respectively, the mean proportion of trials in the first distribution is greater than $1/(K + 1)$, and in fact is

$$\frac{\left(\dfrac{1}{K+1}\right)(K/p)}{\left(\dfrac{1}{K+1}\right)(K/p) + \left(\dfrac{K}{K+1}\right)(1/Kp)} = \frac{K}{K+1}. \tag{11.3.3}$$

The mean waiting time is

$$E_1(n) = \left(\frac{1}{K+1}\right)(K/p) + \left(\frac{K}{K+1}\right)(1/Kp) = 1/p. \tag{11.3.4}$$

Clustering implies an excess of short and long waiting times, the mean affection frequency and waiting time remaining unchanged. On the hypothesis H_1 of clustering, the likelihood is

$$P_1(n) = \left(\frac{1}{K+1}\right)(p/K)(1 - p/K)^{n-1} + \left(\frac{K}{K+1}\right)(Kp)(1 - Kp)^{n-1}. \tag{11.3.5}$$

In practice p is evaluated on the basis of prior experience if the objective is to detect a cluster when it occurs, or empirically from the data being analyzed if prompt detection is less important. The test is sequential if p is specified *a priori*, and a fixed-sample-size test if p is estimated simultaneously. In either case the table of lod scores, $\Sigma \log\{P_1(n)/P_0(n)\}$ as a function of K, gives an appropriate test of clustering, which is reliable in small samples and may be combined and tested for heterogeneity among samples. Table 11.3.1 shows this test applied to categories of spontaneous abortion, for none of which was the clustering significant.

Table 11.3.1: Tests of clustering in a sample of spontaneous abortions

Affection	Sample size	Number affected	$\hat{K}$	$\hat{Z}$
No growth (1) or karyotyped (0)	563	53	1	0
Abnormal (1) or normal (0) | karyotyped	510	243	1.09	.007
Male (1) or female (0) | normal	267	115	1	0
Trisomy (1) or not (0) | abnormal	243	113	1	0
Structural abnormality (1) or not (0) | abnormal	243	12	1	0
Monosomy (1) or not (0) | abnormal	243	61	1.37	.070
Triploidy (1) or not (0) | abnormal	243	40	1.61	.405
Tetraploidy (1) or not (0) | abnormal	243	19	1.89	.392
XXX triploidy (1) or not (0) | abnormal	243	9	2.85	.620

If observations are aggregated into samples whose size is independent of waiting times, the corresponding double binomial distribution for r affected in a sample of size s is appropriate.

$$P_1(r, s) = \left(\frac{K}{K + 1}\right)(p/K)^r(1 - p/K)^{s-r} + \left(\frac{1}{K + 1}\right)(pK)^r(1 - pK)^{s-r}. \tag{11.3.6}$$

If normals are omitted from the sample, this reduces to the double Poisson distribution

$$P_1(r) = \frac{\left(\dfrac{K}{K + 1}\right)(m/K)^r e^{-m/K} + \left(\dfrac{1}{K + 1}\right)(mK)^r e^{-mK}}{r!}.$$

11.4 Clustering in Space

Population structure theory provides methods to test for clustering in space, which may be defined in terms of geography, degree of kinship, hybridity, or any other measure of distance (section 7.7). The null hypothesis of no clustering corresponds to b = 0 in equation 7.7.3.

11.5 Supplementary Tests

When H_0 is accepted or rejected as the result of a sequential test, we should reexamine several points: (a) Was the population at risk correctly determined? A change in birth, death, or migration rate or in administrative boundaries could alter the population. (b) Was the distribution by strata constant? Changes in maternal age at birth or differential fertility of racial or socioeconomic groups could alter affection rates. (c) Was the risk for each group appropriate? In many cases it will be determined from another population which may not be applicable. (d) Did the ascertainment probability change? Relaxation of diagnostic standards to include milder or more questionable cases, standards refined to exclude doubtful cases, splitting or lumping diagnostic categories, earlier diagnosis, and reduced mortality of affected individuals can alter ascertainment probabilities. More energetic screening programs and the establishment of new or better registries have the same effect. Changes in monitor personnel or procedures may be consequential.

The ascertainment probability enters explicitly into surveillance of registries of affected individuals. In principle, changes in ascertainment probability can be detected from the distribution of ascertainments among registered cases. The truncated Poisson distribution arises when there is an indefinitely large number of independent scores, each with a constant ascertainment probability. The truncated Skellam distribution represents a known, limited number of sources, the ascertainment probabilities of which vary as a beta distribution. The truncated negative binomial distribution is a generalization of the Poisson case, applicable to an indefinitely large number of sources whose mean number of ascertainments varies among individuals as a gamma distribution. In practice, one of these formulations is probably adequate, unless the number of sources is as small as two or three. Specific probabilities may be fitted with residual degrees of freedom to test the hypothesis of independence if the number of sources is three or more.

These and other supplementary tests of distributional assumptions and temporal trends should always be made when a sequential test terminates, especially if there is a possibility that a significant health hazard has been detected (or missed).

Some of the advantages of sequential surveillance are found also in sampling inspection schemes that use cumulative sums. These fix the average run length, rather than the type II error, and therefore the properties of the test are less obvious. The calculations are approximate and less flexible, and they make no allowance for changes in sample size, ascertainment probability, or risk factors such as maternal age or birth order that differ among population strata. The practical advantage of one method over the other may be small, but is entirely on the side of sequential tests.

Whenever a surveillance program is initiated, consideration should be given to a sequential design. The principles of a sequential test apply to surveillance, whether the hazard is mutagenic, teratogenic, or carcinogenic. However, heterogeneity of risk categories presents a problem. For example, thalidomide has a specific effect on phocomelia, which would not be detected if cases were pooled with other malformations. One possibility is to conduct a sequential test against pooled categories, with a likelihood ratio examination of specific entities at the termination of the test. Then if m_j is the expected number of the jth category of abnormality under H_0, and n_j is the observed number,

$$2 \sum_{j=1}^{k} n_j \ln\left(\frac{n_j}{m_j}\right) \qquad (n_j, m_j > 0)$$

is a likelihood ratio criterion distributed in large samples as χ^2 with $k - 1$ degrees of freedom under H_0. This test is sensitive to great increase in a rare diagnostic group, with the usual cautions about changes in diagnostic standards and ascertainment probability. Once suspected, a diagnostic category may then be subjected to an appropriate sequential test. The advantages of sequential surveillance are not only that it requires much smaller expected sample sizes for given type I and type II errors. In addition, it provides a surveillance protocol, with sufficiently frequent decisions (to continue sampling, make supplementary tests, and accept or reject H_0) that a registry becomes more than a shoebox full of unanalyzed data. Sequential tests can contribute to the quality of a registry, the morale of its investigators, and the value of a surveillance program to the population being monitored. When surveillance leads to rejection of the null hypothesis, the hazard must be identified and its dose response, mode of action, and means of control determined.

Surveillance for biological hazards is a good illustration of the rule that the first questions to be answered in genetic epidemiology lead to deeper and more interesting ones.

11.6 Questions

1. *Western European countries marketed thalidomide, while the U.S. did not. Predict the response when thalidomide was found to be embryotoxic.*

There was a marked reduction in drugs taken during pregnancy in the involved countries, but not in the U.S.

2. *Efficient systems have been developed to detect mutation and chromosome breakage in cultures of human cells and bacteria. Is this method to screen drugs and environmental agents competitive with epidemiological studies?*

If cell cultures suggest that a hazard may exist, it should be regulated. If cell cultures suggest that a hazard does not exist, mammalian test systems for embryotoxicity may be used. If all tests are negative, the agent may be released for human use. However, epidemiological studies have detected hazards not recognized from animal experiments.

3. *List some mutagens for bacteria in widespread use.*

Caffeine, nitrites, benzene

4. *What potent carcinogens are in widespread use?*

Cigarettes, petroleum and coal products

11.7 Bibliography

Berg K (ed): Genetic Damage in Man Caused by Environmental Agents. Academic Press, New York, 1979

Klingberg MA, Weatherall JAC, Papier C (eds): Epidemiologic Methods for Detection of Teratogens. Karger, Basel, 1979

Morton NE, Lindsten J: Surveillance of Down's syndrome as a paradigm of population monitoring. Hum Hered 26: 360–371, 1976

Wald N.: Sequential Analysis. John Wiley, New York, 1947

Weatherall JAC, Haskey JC: Surveillance of malformations. Br Med Bull 32: 39–44, 1976

Appendix: Statistical Methods

Like other sciences genetic epidemiology advances by testing hypotheses, rejecting those that fail to fit critical observations. A successful hypothesis is never proven, but it beats any hypothesis that must be rejected. The most successful hypotheses, called scientific laws, predict a large number of tested consequences from a few assumptions. Examples are the laws of Mendel and Darwin.

Genetic epidemiology is among the sciences that depend largely on observation, because of the ethical, legal, and practical limitations of human experimentation. Observational sciences like astronomy and geology are no less precise than experimental sciences like chemistry, but they use more elaborate statistical methods to control errors due to the inherent variability of observational data. These methods include both tests of hypotheses and estimates of parameters. Since nonparametric statistics are of little scientific value, genetic epidemiology follows the classical approach through probability and maximum likelihood.

A.1. Elementary Probability

We require only the notions of probability which are familiar to most college students. The basic concept is the *sample space* consisting of all possible events in an idealized experiment, each such event being represented by one *sample point*. A function defined on a sample space is called a *random variable*. Here are the definitions and rules for sample spaces.

The probability P(A) of any event A is the sum of the probabilities of all sample points in it.

$$0 \leq P(A) \leq 1 \tag{A.1.1}$$

Now consider two arbitrary events A_1 and A_2. Since the event A_1 may sometimes include A_2 and vice versa, we have

$$P(A_1 \cup A_2) \leq P(A_1) + P(A_2) \tag{A.1.2}$$

Any point contained both in A_1 and A_2 is counted twice on the right side of equation A.1.2, which therefore exceeds the left side by the amount $P(A_1A_2)$, the probability that both A_1 and A_2 are realized. We therefore deduce

Addition of events. For any two events A_1 and A_2, the probability that either A_1 or A_2 or both occurs is

$$P(A_1 \cup A_2) = P(A_1) + P(A_2) - P(A_1A_2)$$

The generalization of this theorem is obvious. Let

$$S_1 = \sum_i P(A_i)$$
$$S_2 = \sum_{i,j} P(A_iA_j)$$
$$S_3 = \sum_{i,j,k} P(A_iA_jA_k)$$
etc.

The probability of the realization of at least one among the events A_1, A_2, ... A_n is given by

$$P(A_1 \cup A_2 \cup ... \cup A_n) = S_1 - S_2 + S_3 - S_4 + ... \pm S_n \tag{A.1.4}$$

Conditional probabilities are an important tool in genetic epidemiology. We seek the probability of an event A, given that another event H has occurred.

Let H be an event with positive probability. For an arbitrary event A we shall write

$$P(A \mid H) = \frac{P(AH)}{P(H)} \tag{A.1.5}$$

The quantity so defined will be called the conditional probability of A on the hypothesis H (or for given H).

Taking conditional probabilities of various events with respect to a particular hypothesis H amounts to choosing H as a new sample space; all probabilities must be multiplied by the constant factor $1/P(H)$ to reduce the total probability of the sample space to unity. Three extensions of this definition are often used:

$$P(A) = \sum_j P(A \mid H_j) \cdot P(H_j) \tag{A.1.6}$$

$$P(ABC) = P(A \mid BC) \cdot P(B \mid C) \cdot P(C) \tag{A.1.7}$$

$$P(H_j \mid A) = \frac{P(H_j) \cdot P(A \mid H_j)}{\sum\limits_j P(H_j) \cdot P(A \mid H_j)} \qquad \text{(Bayes' theorem)} \qquad\qquad \text{(A.1.8)}$$

As an illustration of these principles, consider the probability that a family with no daughters have only one son. For this problem, we agree that a couple with no children is not to be counted as a family. We let $f(s)$ be the probability that a family picked at random from the population have exactly s children ($s = 1, 2, \ldots$), and we suppose that the probability that a child be a boy is exactly $1/2$. Let A denote the event "the family has boys but no girls." With these conventions, it is obvious from equation A.1.6 that $P(A) = \sum_{s=1}^{\infty} (\frac{1}{2})^s f(s)$. Similarly H_1 is the event "only one child" and AH_1 means "a boy without sibs." Then by equation A.1.8, we have

$$P(H_1 \mid A) = \frac{P(AH_1)}{P(A)} = \frac{(\frac{1}{2})f(1)}{\sum\limits_{s=1}^{\infty} (\frac{1}{2})^s f(s)}$$

as the probability that a family with no daughters have only one son.

The concept of independent events is central to genetics. Under panmixia, autosomal alleles in uniting gametes are independent; at panmictic equilibrium, genes at different loci are independent; the genotypes of different children in a family are independent; unlinked loci segregate independently. Independence is defined as follows.

Two events A and H are said to be stochastically independent (or independent, for short) if

$$P(AH) = P(A) \cdot P(H) \qquad\qquad \text{(A.1.9)}$$

We may now formulate the notion of independent trials.

Let T' and T'' be two sample spaces and denote their points by $A'_1, A'_2, \ldots$ and $A''_1, A''_2, \ldots$. Let the corresponding probabilities be $q'_1, q'_2, \ldots$ and $q''_1, q''_2, \ldots$. The succession of the two experiments is described by the space with points (A'_i, A''_j). Saying that the two successive experiments are independent means defining probabilities by

$$P(A'_i, A''_j) = q'_i q''_j \qquad\qquad \text{(A.1.10)}$$

More generally, if $A_1, A_2, \ldots A_N$ represent trials from N sample spaces (not necessarily different), the trials are said to be independent if

$$P(A_1 A_2 \ldots A_N) = P(A_1) \cdot P(A_2) \ldots P(A_N)$$

This leads us to the concept of Bernoulli trials.

Repeated independent trials are called Bernoulli trials, if there are only two possible outcomes for each trial and their probabilities (q, 1 − q) remain the same throughout the trials. (A.1.11)

Repeated random drawings from a gene pool containing at each drawing R gametes bearing the gene A, and N–R gametes bearing different alleles, represent Bernoulli trials with q = R/N. It is evident that the gene pool must be indefinitely large (N → ∞) to meet the condition that R and N–R remain constant. We may think of a pool of all gametes in virtually infinite numbers, from which gametes are drawn at random. Alternatively, we may prefer to think of a gamete being returned to the gene pool after each sampling, so that the pool remains constant: This is called *sampling with replacement*.

There are s(s − 1) ... (1) ≡ s! possible orderings (permutations) of s objects, and, therefore, r! (s − r)! permutations of two sets of size r and s–r. Elementary probability theory proves the theorem:

Let P(r; s, q) be the probability that Bernoulli trials with probability q for success and 1 − q for failure result in r successes and s − r failures (0 ≤ r ≤ s). Then

$$P(r; s, q) = \binom{s}{r} q^r (1 - q)^{s-r} \tag{A.1.12}$$

where

$$\binom{s}{r} = \frac{s!}{r!\,(s - r)!} \text{ and } s! = s(s - 1) \ldots (1).$$

In particular, the probability of no successes is $(1 - q)^s$, and of at least one success is $1 - (1 - q)^s$.

The number of successes r is a random variable, whose distribution P(r; s, q) is called the *binomial distribution*. Its generalization is the *multinomial distribution*.

Let there be s independent trials, where each trial can have one of several outcomes. Denote the possible outcomes of each trial by $A_1, A_2, \ldots, A_m$, and suppose that the probability of the realization of A_i in each trial is q_i (i = 1, 2, ... m). The probability that in s trials A_1 occurs r_1 times, A_2 occurs r_2 times, etc. is

$$P(r_1, r_2, \ldots r_k; s, q_i) = \frac{s!}{r_1!\, r_2! \ldots r_m!}\, q_1^{r_1} q_2^{r_2} \ldots q_m^{r_m}$$

where

$$\sum_{i=1}^{m} q_i = 1 \quad \text{and} \quad \sum_{i=1}^{m} r_i = s$$

For $m = 2$, this reduces to Bernoulli trials.

Not all distributions in genetics assume independent trials. The most important exception is *sampling without replacement*. In a population of S elements there are R_1 of type 1, R_2 of type 2, ..., R_m of type m. Suppose a sample of s elements is chosen at random, without replacement. Then the population size S is reduced successively to S-1, S-2, ..., S-s.

The hypergeometric distribution $P(r_1, r_2, \ldots r_m ; S, R_i)$ is the probability of obtaining $r_1, r_2, \ldots r_m$ elements in sampling without replacement, or

$$P(r_1, r_2, \ldots r_m ; S, R_i) = \frac{\binom{R_1}{r_1}\binom{R_2}{r_2}\cdots\binom{R_m}{r_m}}{\binom{S}{s}}$$

where $\sum R_i = S$ and $\sum r_i = s$.

It is obvious that if S, R_i approach infinity, the probability of the ith element remains constant during sampling, $q_i = R_i/S$. Since

$$\lim_{R_i \to \infty} \binom{R_i}{r_i} = R_i^{r_i},$$

we conclude that as S, R_i increase without limit, the hypergeometric distribution converges to the multinomial,

$$P(r_1, r_2, \ldots r_m ; S, R_i) \to P(r_1, r_2, \ldots r_m ; s, q_i). \tag{A.1.15}$$

The multinomial distribution itself approaches a limit as the sample size increases. We state without proof: as s increases without limit, the multinomial distribution converges to the $m - 1$ dimensional normal distribution,

$$f(\hat{q}_1, \hat{q}_2, \ldots \hat{q}_{m-1} ; K_{ij}, q_i) = Ce^{-(\frac{1}{2})\Sigma(\hat{q}_i - q_i)(\hat{q}_j - q_j)K_{ij}} \tag{A.1.16}$$

where $q_i = R_i/S$, $\hat{q}_i = r_i/s$, C is a constant dependent on m but on the q_i, and K_{ij} is the element in the ith row and j^{th} column of the *information matrix* (section A.4.5).

To illustrate the multinomial distribution, consider genotype and mating type frequencies for an autosomal system under panmixia. For genotype frequencies, we agree to make no distinction between A^iA^j and A^jA^i; we do not care which allele came from the father and which from the mother. From equation A.1.12 we have

$$P(A^iA^i) = p_{ii} = \binom{2}{2}q_i^2 = q_i^2$$

$$P(A^iA^j) = 2p_{ij} = \binom{2}{1}q_i q_j = 2q_i q_j \qquad (i \neq j)$$

For mating type frequencies (with reciprocal matings pooled), the elementary genotype probabilities enter into equation A.1.12 with double subscripts:

$$P(A^iA^j \times A^iA^j) = m_{(ij)(ij)} = \binom{2}{2}p_{ij}^2 = p_{ij}^2$$

$$P(A^iA^j \times A^kA^l) = 2m_{(ij)(kl)} = \binom{2}{1}p_{ij} p_{kl} = 2p_{ij} p_{kl} \qquad (i, j \neq k, l).$$

A.2. The Principle of Maximum Likelihood

Given a sample from a population with m alleles and $p_{ij} = q_i q_j$, intuition may suggest different procedures to estimate the vector of gene frequencies q. For example, consider the MN factors:

Genotype	MM	MN	NN	
Factors	10	11	01	
Expected proportion	q^2	$2q(1 - q)$	$(1 - q)^2$	
Observed number	r_1	r_2	r_3	$(r_1 + r_2 + r_3 = s)$

It is clear that we could estimate q by $\sqrt{r_1/s}$, since q^2 is the expected value of r_1/s. We could also estimate q by $(2r_1 + r_2)/2s$, corresponding to a count of 2 M genes for each MM person and 1 M gene for each MN person. Readers with mathematical intuition and/or experience will recognize that by combining r_1, r_2, and r_3 in various ways we can devise an indefinitely large number of estimation procedures, all of which are sensible in that they converge to the true value as s increases. For example, we might estimate

$$q^2 = r_1/s, \qquad q = \sqrt{r_1/s}$$

or

$$(1 - q)^2 = r_3/s, \qquad q = 1 - \sqrt{r_3/s},$$

or combine these estimates in proportion θ,

$$q = \theta \sqrt{r_1/s} + (1 - \theta)(1 - \sqrt{r_3/s}),$$

where $0 \leq \theta \leq 1$.
Again, we could set

$$\frac{[2q(1 - q)]^2}{(1 - q)^2} = 4q^2 = \frac{r_2^2}{r_3}, q = \frac{r_2}{2\sqrt{r_3}}$$

or $\quad \dfrac{[2q(1 - q)]^2}{q^2} = 4(1 - q)^2 = \dfrac{r_2^2}{r_1}, q = 1 - \dfrac{r_2}{2\sqrt{r_1}},$

or combine these estimates in proportion θ,

$$q = \theta \frac{r_2}{2\sqrt{r_3}} + (1 - \theta)\left[1 - \frac{r_2}{2\sqrt{r_1}}\right].$$

The method of *maximum likelihood* avoids this infinity of estimates by choosing among all specified alternatives, the one with the highest probability. Thus in a mating known to be Aa × Aa we would guess that an A- child is Aa, since this is twice as likely as AA, ie,

$$P(Aa \mid A\text{-}) = \frac{P(Aa)}{P(A\text{-})} = \frac{1/2}{3/4} = 2/3$$

Similarly, in the absence of other information we would agree that the most likely Rh genotype with the phenotype CcDEe is CDe/cDE, since the other alleles are much rarer. Applied to the problem of estimation, where the alternatives are continuous, the principle of maximum likelihood may be expressed in this way: take as the estimates the values that maximize the probability of the observations.

We may illustrate this by the univariate normal distribution (eq A.1.16), which is

$$f(\hat{q}; \sigma^2, q) = Ce^{-(\hat{q}-q)^2/2\sigma^2}. \tag{A.2.1}$$

Clearly this value cannot be greater than C, and it takes this maximum when $\hat{q} = q$. Thus, the maximum likelihood (ML) estimate of the population mean q is the sample mean $\hat{q}$.

It is not so easy to show that the ML estimate and the sample means coincide for the multinomial distribution. Consider the simplest case of Bernoulli trials with probability q for success and $1 - q$ for failure, resulting in r successes and $s - r$ failures. The probability of the observations (eq A.1.12) is

$$P(r; s, q) = \binom{s}{r} q^r (1 - q)^{s-r}.$$

How can we find the estimate of q which maximizes this probability? Actually, we shall see that it is simpler to maximize the logarithm of the probability. Since independent probabilities are multiplicative, their logarithms are additive. We define *logarithmic likelihood* L as the natural logarithm of the probability, omitting any constant multiplier. Then if L_i is so defined for the ith sample, the logarithmic likelihood of a set of independent samples is

$$L = \Sigma \, L_i. \qquad\qquad\qquad\qquad\qquad (A.2.2)$$

This additivity is helpful in applications, for which we require some elements of differential calculus.

A.3. Differential Calculus

Consider a function y of x,

$$y = f(x).$$

Differential calculus is concerned with the rate of change of y with respect to x, which is also the tangent to y. This is symbolized by $\partial y / \partial x$, usually called the partial derivative of y with respect to x. It is obtained as the limit of

$$\frac{f(x + \Delta) - f(x)}{\Delta}$$

as Δ becomes infinitesimally small.
    Suppose $y = x^n$. Then

$$\partial y / \partial x = \lim_{\Delta \to \infty} \left[\frac{(x + \Delta)^n - x^n}{\Delta} \right] = \lim \left[\frac{nx^{n-1}\Delta + \binom{n}{2}x^{n-2}\Delta^2 + \cdots + \Delta^n}{\Delta} \right]$$

$$= nx^{n-1}.$$

Several other simple formulae from differential calculus, derived in the same way, are frequently required in applications of maximum likelihood. They are:

Theorem 1. If $y = ax^n$, where a is a constant, then $\partial y/\partial x = anx^{n-1}$
Theorem 2. If $y = e^u$, where u is a function of x, then $\partial y/\partial x = e^u(\partial u/\partial x)$
Theorem 3. If $y = uv$, where u and v are functions of x, then $\partial y/\partial x$
$\qquad\qquad = u(\partial v/\partial x) + v(\partial u/\partial x)$
Theorem 4. If $y = u/v$, then $\partial y/\partial x = \{v(\partial u/\partial x) - u(\partial v/\partial x)\}/v^2$
Theorem 5. If $y = \ln u$, then $\partial y/\partial x = (\partial u/\partial x)/u$

These theorems may be used to find the maximum value of a probability function. Since $\partial y/\partial x$ is the tangent to y, if $\partial y/\partial x = 0$ then y must have a maximum or minimum at that point. As a corollary, we see by theorem 2 that since $\partial e^u/\partial x = e^u(\partial u/\partial x)$, then $\partial e^u/\partial x = 0$ implies $\partial u/\partial x = 0$, so the maximum or minimum of a function occurs at the same point as the maximum or minimum of the logarithm of that function.

We shall need one other result from differential calculus. If $z = f(x, y)$ is a function of two variables, and $y = f(x)$, we may define derivatives dz/dx and dz/dy as if x and y were not functionally related. To return to the partial derivative $\partial z/\partial x$, which takes this relationship into account, we have

$$\frac{\partial z}{\partial x} = \frac{dz}{dx} + \frac{dz}{dy} \cdot \frac{\partial y}{\partial x} \qquad\qquad (A.3.1)$$

Nearly all sampling distributions encountered in nature are such that there is at most a single permissible value at which the derivative equals 0, and at this point the probability is a maximum. However, sometimes there is no point at which $\partial y/\partial x = 0$. Then the ML estimate is simply the value for which the probability P is a relative maximum over the set of possible values of the parameters.

For example, consider the binomial distribution

$$P(r: s, q) = \binom{s}{r} q^r(1 - q)^{s-r} \qquad\qquad (A.3.2)$$

$$L = r\ln q + (s - r)\ln(1 - q)$$

$$\frac{\partial L}{\partial q} = r/q - (s - r)/(1 - q).$$

Setting this equal to 0, we obtain as the ML solution

$$\hat{q} = r/s,$$

which shows that the sample proportion is the ML estimate of the true value q. We say that q is a parameter, and $\hat{q}$ is its ML estimate. Now, however, suppose that r = 0. Then

$$\partial L/\partial q = -s/(1 - q).$$

Since q must lie between 0 and 1, there is no permissible value of q for which $\partial L/\partial q = 0$. The closest $\partial L/\partial q$ can get to 0 is for $\hat{q} = 0$. Thus the ML estimate of the probability of an event which is not observed is the sample proportion, or 0, but this is not a root of the equation $\partial L/\partial q = 0$.

A.4. Properties of Maximum Likelihood Estimates

Although the principle of maximum likelihood has a strong intuitive appeal, its justification rests on more solid grounds. R. A. Fisher and his successors showed that in large samples the ML theory has optimum statistical properties. To understand these, it is necessary to consider what we mean by a "good" estimate.

The estimation problem may be stated as follows: one is investigating a population with a distribution $f(x; q_i)$, where x is the random variable and q_1, q_2, ... q_n are parameters in the distribution. On the basis of a random sample of observations, we wish to find functions of the observations called *estimates* and represented by $\hat{q}_i$ for i = 1, 2, ... n, such that the distribution of these functions will be concentrated as closely as possible near the true values of the parameters.

Consider for simplicity the case of a single parameter, q. Suppose $\hat{q}_1$, $\hat{q}_2$, are different estimates of q. One method of comparing two estimates is by their relative efficiency, defined in terms of mathematical expectation.

By E(x), the *mathematical expectation* of a random variable x, we mean $E(x) = \Sigma x P(x)$, where P(x) is the distribution of x $[0 \leq P(x), \Sigma P(x) = 1]$.

If $E(\hat{q}_1 - q)^2 = \sigma_1^2$ and $E(\hat{q}_2 - q)^2 = \sigma_2^2$, then the efficiency of $\hat{q}_1$ relative to $\hat{q}_2$ is defined as σ_2^2/σ_1^2; this is usually expressed as a percentage. If the efficiency of $\hat{q}_1$ relative to $\hat{q}_2$ is greater than 100 percent, $\hat{q}_1$ may reasonably be regarded as a better estimate than $\hat{q}_2$.

The quantity $E(\hat{q} - q)^2$ is called the *variance* of the estimate $\hat{q}$ of parameter q.   (A.4.1)

We are now ready to state some of the properties of ML theory:

Property 1.   *Unbiasedness.* The ML estimate converges in large samples to the value of the parameter, ie, if q̂ is an ML estimate of a parameter q in a sample of size s,

$$E(\hat{q}) = q \quad \text{as} \quad s \to \infty$$

Property 2.   *Consistency.* The variance of an ML estimate approaches zero as the sample size becomes indefinitely large, ie,

$$E(\hat{q} - q)^2 \to 0 \quad \text{as} \quad s \to \infty$$

Property 3.   *Efficiency.* In large samples, no consistent estimate has a smaller variance than the ML estimate. Therefore no other estimating procedure is better in large samples. If q̂ is the ML estimate and q̂′ another consistent estimate,

$$E(\hat{q} - q)^2 \leq E(\hat{q}' - q)^2 \quad \text{as} \quad s \to \infty$$

Property 4.   *Invariance.* The ML estimate is invariant, in the sense that if q̂ is the ML estimate of q, then f(q̂) is the ML estimate of a function f(q).

Property 5.   *Sufficiency.* An estimate is said to be sufficient if its distribution is independent of the true value of the parameter. Sufficient statistics contain all the information that a sample can give about a parameter. If a sufficient estimate exists, the ML estimate is sufficient.

Property 6.   *Minimum confidence interval.* For every sample there is a range of values of the parameter q for which the deviation of the estimate q̂ from q is said to be nonsignificant. This range is called a confidence interval. If the interval includes all the values of the parameter from which the sample does not deviate at the .05 level of significance, the interval is called a 95 percent confidence interval. If this significance level is adhered to, in the long run the parameter will be correctly asserted to lie within the interval in 95 percent of the cases, and falsely asserted to lie within the interval in 5 percent of the cases. The statement that a confidence interval does not include some value q′ is equivalent to the statement that the deviation of q̂ from q′ is significant. In large samples no other estimating procedure can give a smaller confidence interval than ML theory, according to which a confidence interval in large samples takes the form

$$\hat{q} - t\sigma < q < \hat{q} + t\sigma$$

where σ is the square root of the variance of $\hat{q}$ and is called the *standard error* of $\hat{q}$, and t takes the following values for different confidence levels $1 - \alpha$:

$1 - \alpha$	t
.90	1.645
.95	1.960
.99	2.576

(A.4.2)

Property 7. *Estimation of the variance* σ^2. The usefulness of the ML method is not limited to estimation of parameters; it also gives the approximate variance of the estimate, as a measure of its precision. Theorem A.1.16 is a special case of the *central limit theorem*, according to which any distribution with finite means and variances converges in large samples to the normal distribution. Consider for simplicity the univariate case,

$$f(\hat{q}; \sigma^2, q) = Ce^{-(\hat{q}-q)^2/2\sigma^2}$$

We saw in equation A.2.1 that the ML estimate of the population mean q is the sample mean $\hat{q}$. We may verify this by the calculus

$$L = -(\hat{q} - q)^2/2\sigma^2$$

$$\frac{\partial L}{\partial q} = \frac{\hat{q} - q}{\sigma^2}$$

(A.4.3)

Equating $\partial L/\partial q$ to zero, we confirm that $\hat{q}$ is indeed the ML estimate of q.

Differentiating $\dfrac{\partial L}{\partial q}$ we obtain

$$\frac{\partial^2 L}{\partial q^2} = -1/\sigma^2.$$

The amount of information about q is defined as

$$K_{qq} = -\frac{\partial^2 L}{\partial q^2} = 1/\sigma^2 \qquad (A.4.4)$$

Property 8. *Simultaneous estimation of two or more parameters.* Everything that has been said about the univariate case is easily generalized to cover n linearly independent parameters, $q_1, \ldots, q_n$. By linear independence we mean that none of the parameters is a linear function of the others. The n-variate normal distribution (equation A.1.16) is Ce^L, where

$$L = \left(-\frac{1}{2}\right) \sum_{i,j=1}^{n} (\hat{q}_i - q_i)(\hat{q}_j - q_j)K_{ij}$$

The ML equations are

$$\frac{\partial L}{\partial q_i} = 0, \qquad (i = 1, 2, \ldots n)$$

and the elements of the *information matrix* are

$$-\frac{\partial^2 L}{\partial q_i \partial q_j} = K_{ij} \qquad (A.4.5)$$

The inverse of K is called the *covariance matrix*, the main diagonal of which gives the variances of the estimates $\hat{q}_i$.

Often we are interested not in pure estimation but in testing some hypothesis about the parameters q_i. This is called the *null hypothesis* H_0. In the theory of large samples the quantity $-2L$ is said to be distributed as χ^2 with η *degrees of freedom* on the null hypothesis H_0 that specifies the value of η independent parameters, all parameters neither estimated nor tested being omitted from L (table A.4.1). This is called the *likelihood ratio* test. In particular with $\eta = 1$, $-2L$ becomes $(\hat{q} - q_0)^2/\sigma^2$, where q_0 is the value under H_0 and $\hat{q}$ is an ML estimate with variance σ^2. Large values of $-2L$ are improbable on the null hypothesis. For example, with $\eta = 1$ the value of 3.84 is expected to be exceeded with probability .05 under H_0 (table A.4.1).

An alternative test is available for multinomial distributions. If the null hypothesis specifying η independent parameters gives expected numbers e_i $(i = 1, \ldots M)$, and the observed numbers are o_i, then

$$\chi^2 = \sum \frac{(o_i - e_i)^2}{e_i} = \sum \frac{o_i^2}{e_i} - N, \qquad (A.4.6)$$

where χ^2 has $M - \eta - 1$ degrees of freedom and $N = \Sigma_i o_i = \Sigma_i e_i$ is the total number of observations. The corresponding likelihood ratio test is

$$-2L = 2\Sigma \; o_i \ln(o_i/e_i), \qquad (o_i > 0). \tag{A.4.7}$$

The two tests are equivalent in the limit for large samples, but may give different results in small samples. If, in a particular application, the two tests are in close agreement, this increases our confidence in large-sample theory.

Table A.4.1: Significant values of χ^2

Degrees of freedom	Significance level	
	.05	.01
1	3.84	6.64
2	5.99	9.21
3	7.82	11.34
5	11.07	15.09
10	18.31	23.21
15	25.00	30.58
30	43.77	50.89

The credentials of ML theory are impressive, but the reader should be aware of an important restriction: the optimum properties are realized only in indefinitely large samples, and the central limit theorem may be unreliable in small samples. We shall meet such a case in studying linkage. In practice ML theory seems always to give good estimates, but the confidence intervals and significance tests are approximate. There is seldom a preferable alternative in large samples or when two or more parameters are to be estimated simultaneously, but more accurate confidence intervals and significance tests have been devised for the case of a single parameter in small samples.

Letting s^2 be an estimate of variance based on η degrees of freedom, where the $E(s^2) = \sigma^2$, then χ^2 is of the form $\eta s^2/\sigma^2$. A related statistic is

$$F = s_1^2/s_2^2 \tag{A.4.8}$$

where s_1^2, s_2^2 are two independent estimates of the same variance under some null hypothesis, based on η_1 and η_2 degrees of freedom, respectively. Fisher derived the distribution of F for a normally distributed random variable, making it unnecessary to know σ^2 exactly as the χ^2 test assumes.

A test that does not depend on normality assumes a probability distribution $P(x; \theta)$, where x is a random variable and θ is a single parameter. Then $\theta_1 < \theta < \theta_2$ is a confidence interval of strength at least $1/A$, where $P(x; \theta_1) = P(x; \theta_2) = p(x; \theta)/A$ for $A > 0$, as is proven in equation A.8.5. This serves both to test the null hypothesis that $\theta' = \theta_0$ and to provide a confidence interval for θ in samples of any fixed size.

A.5 Boolean and Matrix Algebra

Genetic epidemiology requires elements of Boolean and matrix algebra which may be unfamiliar to the reader. Fortunately a few pages can give all the background required.

Consider the binary numbers (0,1), signifying absence and presence of an attribute, respectively. There are three elementary operations on these numbers in Boolean algebra.
1. *Complementation* ($\sim$)

$$\sim 1 = 0 \tag{A.5.1}$$
$$\sim 0 = 1$$

This may be interpreted as "not 1 signifies 0, and vice versa."
2. *Union* $(\cup)$

$$1 \cup 1 = 1 \cup 0 = 0 \cup 1 = 1 \tag{A.5.2}$$
$$0 \cup 0 = 0$$

This may be interpreted as "1 unless both are 0."
3. *Intersection* $(\cap)$

$$1 \cap 1 = 1 \tag{A.5.3}$$
$$1 \cap 0 = 0 \cap 1 = 0 \cap 0 = 0$$

This may be interpreted as "not 1 unless both are 1." It will be noticed that union corresponds to dominance and intersection to recessivity or other interaction. In a factor-union system neither complementation nor intersection is required (or permitted) to generate phenotypes, but both complementation and intersection are used to exclude parentage.

Application of Boolean algebra to several numbers is straightforward, since each is considered separately. It is convenient to tabulate the numbers in a row, called a *row vector*, the number of entries being termed the *order* n of the vector. For example, 101100 and 011010 are two binary vectors of order 6. The Boolean operations are defined only on vectors of the same order, as in the following examples:

$$101100 \sim 010011$$

$$101100 \cup 011010 = 111110$$

$$101100 \cap 011010 = 001000$$

Vectors will be denoted by small letters. It is conventional to bracket a vector, but we shall omit brackets when dealing with binary row vectors. To designate a column vector we shall add a prime to the corresponding row vector. Thus, if x = 001,

$$x' = \begin{bmatrix} 0 \\ 0 \\ 1 \end{bmatrix}.$$

The change of a row vector to a column vector, or conversely, is called *transposition*. Then x is the transpose of x', and vice versa.

It is often convenient to write vectors in a rectangular array, called a matrix, whose order is the number of rows and columns. Thus

$$\begin{bmatrix} 001 \\ 010 \end{bmatrix}$$

is a binary matrix of order 2 × 3. Matrices will be designated by capital letters and the arrays bracketed, as in the above example. The element a_{ij} of a matrix is the entry in the ith row and j^{th} column.

Transposition of a matrix signifies interchange of rows and columns, and is indicated by a prime. Thus if

$$A = \begin{bmatrix} 001 \\ 010 \end{bmatrix}, \quad \text{then} \quad A' = \begin{bmatrix} 00 \\ 01 \\ 10 \end{bmatrix}.$$

An important class of matrices is the binary, square *unit matrix* I, which has 1's only on the principal diagonal. For example, the unit matrix of order 3 is

$$I = \begin{bmatrix} 100 \\ 010 \\ 001 \end{bmatrix}.$$

The unit matrix plays the same role in matrix algebra as unity plays in ordinary algebra. In fact, ordinary algebra is the special case of order 1 of matrix algebra. A matrix of order 1 is called a *scalar*. The unit matrix is an example of a *symmetrical* matrix, where $a_{ij} = a_{ji}$ for all elements i and j, and therefore $A = A'$.

Any matrix may be partitioned into submatrices, separated by dotted lines. Thus

$$A = \begin{bmatrix} 00 & : & 11 \\ 01 & : & 00 \end{bmatrix}$$

is a *partitioned* matrix. If we let

$$B = \begin{bmatrix} 00 \\ 01 \end{bmatrix} \quad \text{and} \quad C = \begin{bmatrix} 11 \\ 00 \end{bmatrix},$$

then we may compactly write $A = [B : C]$.

To apply ML theory to simultaneous estimation of two or more parameters, we must go beyond Boolean algebra and pass to the general matrix as any rectangular array of quantities. For example,

$$\begin{bmatrix} 3 & \log x \\ e^{kx} & f(y) \end{bmatrix}$$

is a square matrix of order 2. The four elementary operations are generalized in matrix algebra.

1. *Addition*. The sum of two matrices is the matrix of the sums of corresponding elements.

$$\begin{bmatrix} a & b \\ c & d \end{bmatrix} + \begin{bmatrix} e & f \\ g & h \end{bmatrix} = \begin{bmatrix} a+e & b+f \\ c+g & d+h \end{bmatrix}$$

2. *Subtraction*. This is similarly defined.

$$\begin{bmatrix} a & b \\ c & d \end{bmatrix} - \begin{bmatrix} e & f \\ g & h \end{bmatrix} = \begin{bmatrix} a-e & b-f \\ c-g & d-h \end{bmatrix}$$

3. *Multiplication*. The product of two matrices is defined as follows: The element in the ith row and j^{th} column of the product matrix is obtained by multiplying the elements of the ith row of the left-hand matrix by the corresponding elements of the jth column of the right-hand matrix and adding the results.

$$\begin{bmatrix} a & b \\ c & d \end{bmatrix} \cdot \begin{bmatrix} e & f \\ g & h \end{bmatrix} = \begin{bmatrix} ae+bg & af+bh \\ ce+dg & cf+dh \end{bmatrix}$$

Note that multiplication is not commutative: unless A and B are both symmetrical, AB ≠ BA. It should be clear that two matrices cannot be multiplied to give a product AB unless the number of columns in the first matrix A equals the number of rows in the second matrix B.

4. *Division.* As in ordinary algebra, the division of a matrix A by another B is accomplished by multiplying A times the reciprocal of B. The reciprocal of a matrix is called its *inverse*, and is denoted by the superscript -1. Thus B^{-1} is the inverse of B, and conversely. The elements of an inverse matrix will be superscripted, as

$$[b^{ij}] = [b_{ij}]^{-1}$$

The inversion of a large matrix is tedious and in practice is left to a computer program that uses some rapid but rather complicated procedure, the effect of which is as follows. Let A. represent the *determinant* of a square matrix A of order k, and let A_{ij} be the *cofactor* of any element a_{ij}, defined as the determinant of order $k-1$ formed by omitting the ith row and j^{th} column of A, multiplied by $(-1)^{i+j}$. In the elementary theory of determinants, it is shown that the value of a determinant may be obtained by adding the products of the elements of any row by their cofactors, ie,

$$A. = \sum_{j=1}^{k} a_{ij} A_{ij},$$

where any value of i may be used. Thus any determinant may be reduced to combinations of determinants of order 2. For example,

$$\text{if} \quad A = \begin{bmatrix} a & b & c \\ d & e & f \\ g & h & i \end{bmatrix}$$

then $A = a(ei - fh) - b(di - fg) + c(dh - eg)$.

The elements of the *inverse* of a matrix are the transpose of the ratio of the corresponding cofactors to the determinant,

$$a^{ij} = A_{ji}/A. \quad (A. > 0)$$

To justify this definition, we note without formal proof that the product of a matrix and its inverse is the unit matrix, just as $c(1/c) = 1$ in ordinary algebra.

The reader should verify that,

if $\quad A = \begin{bmatrix} 3 & 1 & 1 \\ 2 & 2 & 1 \\ 2 & 2 & 3 \end{bmatrix}$

then $A. = 8$,

$$A^{-1} = \begin{bmatrix} 4/8 & -1/8 & -1/8 \\ -4/8 & 7/8 & -1/8 \\ 0 & -4/8 & 4/8 \end{bmatrix},$$

and $AA^{-1} = A^{-1}A = I$.

Matrices are commonly used in the solution of linear equations. For example, consider the simultaneous equations

$a_{11}x_1 + a_{12}x_2 = c_1.$

$a_{21}x_1 + a_{22}x_2 = c_2.$

This is written in matrix algebra as

$Ax' = c'$,

where

$$A = \begin{bmatrix} a_{11} & a_{12} \\ a_{21} & a_{22} \end{bmatrix}, \quad x' = \begin{bmatrix} x_1 \\ x_2 \end{bmatrix}, \quad \text{and} \quad c' = \begin{bmatrix} c_1 \\ c_2 \end{bmatrix}$$

The reader may easily verify that

$A. = a_{11}a_{22} - a_{12}a_{21},$

$$A^{-1} = \begin{bmatrix} a_{22}/A. & -a_{12}/A. \\ -a_{21}/A. & a_{11}/A. \end{bmatrix},$$

and the solution is

$x' = A^{-1}c'$,

which may be written less succinctly as

$x_1 = (a_{22}c_1 - a_{12}c_2)/A.$

$x_2 = (-a_{21}c_1 + a_{11}c_2)/A.$

The results extend to any number of parameters and necessarily agree with the old-fashioned method of elimination. For example, if we multiply the equations by a_{22}, a_{12}, respectively, and then subtract, we obtain

$$a_{11}a_{22}x_1 + a_{12}a_{22}x_2 = a_{22}c_1$$

$$a_{12}a_{21}x_1 + a_{12}a_{22}x_2 = a_{12}c_2$$

$$(a_{11}a_{22} - a_{12}a_{21})x_1 = a_{22}c_1 - a_{12}c_2$$

or

$$x_1 = (a_{22}c_1 - a_{12}c_2)/A.$$

If a matrix is symmetric, its inverse is also symmetric, so that we need not compute or write the elements above (or below) the principal diagonal nor transpose the cofactor matrix. Thus if

$$A = \begin{bmatrix} 3 & 1 \\ & 2 \end{bmatrix}, \quad \text{then} \quad A. = 5 \quad \text{and} \quad A^{-1} = \begin{bmatrix} 2/5 & -1/5 \\ & 3/5 \end{bmatrix}$$

A.6 Iterative Minimization

For normally distributed random variables,

$$L = -\chi^2/2 = -ns^2/2\sigma^2. \tag{A.6.1}$$

Maximization of the likelihood is equivalent to minimization of $-L$. The solution often cannot be obtained explicitly, but only by some iterative procedure in which successive approximations converge to ML estimates. The classical method is Newton-Raphson iteration.

Any function $u(\hat{q})$, continuous and having derivatives, may be expanded into a *Taylor's series*,

$$u(\hat{q}) = u(q) + \frac{\partial u(q)}{\partial q}(\hat{q} - q) + \ldots\ldots \tag{A.6.2}$$

where q, an estimate of $\hat{q}$, may be any quantity for which the derivatives are finite. If $q = 0$, this is called *Maclaurin's series*. Provided that $|\hat{q} - q|$ is small, the remainder on the right of equation A.6.2 is negligible. This suggests Newton-Raphson iteration, in which the root of an equation

$$u(\hat{q}) = 0$$

is sought by successive approximations,

$$q^{(n)} = q^{(n-1)} + U/K \qquad (A.6.3)$$

where

$$U = u(q)$$

$$K = -\frac{\partial u(q)}{\partial q}$$

are both evaluated at $q^{(n-1)}$.

This process will converge under conditions studied in numerical analysis, which are generally met if $|\hat{q} - q|/\hat{q}$ is small. In the neighborhood of a root convergence is quadratic, the absolute error in the nth iteration being less than or equal to the square of the preceding error,

$$|\hat{q} - q^{(n)}| \le (\hat{q} - q^{n-1})^2$$

For simultaneous equations with two or more parameters, the matrix analog of equation A.6.3 is

$$q^{(n)} = q^{(n-1)} + uK^{-1}, \qquad (A.6.4)$$

where q is the vector of estimates, u is the corresponding vector of functions, and K is the matrix

$$\left[-\frac{\partial u\,(q^{(n-1)})}{\partial q^{(n-1)}} \right].$$

To relate Newton-Raphson iteration to ML estimation, consider first a single parameter q. It is helpful to define

$$u_m = \partial L_m/\partial q \qquad (A.6.5)$$

for the mth observation with logarithmic likelihood L_m as the *ML score* with respect to the parameter q. Then to the logarithmic likelihood for n independent items of data

$$L = \sum_{m=1}^{n} L_m$$

there corresponds a total score

$$U = \sum_{m=1}^{n} u_m.$$

In large-sample theory,

$$L = -(\hat{q} - q)^2/2\sigma^2$$

$$U = \partial L/\partial q = \frac{\hat{q} - q}{\sigma^2}$$

and the amount of information is the variance of U,

$$K = E(U^2) = \frac{E(\hat{q} - q)^2}{\sigma^4} = 1/\sigma^2, \qquad (A.6.6)$$

in agreement with equation 4.3.4. Since

$$E(U^2) = \sum_{m=1}^{n} \left[E\left(u_m^2\right) \right]$$

we define

$$k_m = E(u_m^2) \qquad (A.6.7)$$

as the amount of information contributed by the mth observation.

Application to Newton-Raphson iteration is straightforward, substituting U for u(q) and combining equation A.4.4, A.6.6, and A.6.7 to give

$$K = -E\left(\frac{\partial U}{\partial q}\right) = E(U^2) = \Sigma k. \qquad (A.6.8)$$

It has been suggested that replacement of $\partial U/\partial q$ by its expected value may impair convergence, but this has not been shown to be of practical importance. Equation A.6.8 allows the same calculation to provide $\hat{q}$ and its variance.

To estimate a vector q with two or more linearly independent parameters we use equation 4.5.3 with the vector of scores

$$u = \left(\frac{\partial L}{\partial q}\right)$$

and the information matrix

$$K = [E(U_i U_j)] = [\Sigma k_{ij}] \qquad (A.6.9)$$

where $k_{ij} = E(u_i u_j)$ for parameters q_i, q_j.

In large sample theory the quantity

$$-2L = uK^{-1}u',$$
(A.6.10)

is distributed as χ^2 with η degrees of freedom under H_0, a null hypothesis about η parameters. Under a model containing ρ linearly independent parameters, of which η are specified by the null hypothesis, the *order* is ρ, the degrees of freedom are η, and the *rank* $\rho - \eta$ is the number of parameters estimated.

For convergence $q^{(0)}$ must be sufficiently close to $\hat{q}$, with other conditions not readily demonstrated in actual data except by trial and error. Therefore, we have no guarantee that iteration will converge, and several choices of $q^{(0)}$ may sometimes have to be made, especially in the multiparameter case, using either intuition or an inefficient estimate. If convergence is still not obtained, it is usually necessary to simplify the model, for example by assuming that an idiomorph known to exist, but not demonstrated in the sample, is absent ($q = 0$). This is called "reducing the rank" of the model.

In the univariate case, we know that if U changes sign between two values of q, say q' and q'', then the ML root $\hat{q}$ lies in that interval and may be estimated by

$$q \doteq q' + \frac{U'(q'' - q')}{|U'| + |U''|}$$
(A.6.11)

This linear interpolation is often helpful in finding a better value of $q^{(0)}$. Unfortunately in the multiparameter case we cannot be sure that a root lies in the interval, because other estimates are changing simultaneously.

There is no theory that assures in small samples that the root of $U = 0$ may not be a minimum of L, a saddle point, or a local maximum. The reader should be aware of this possibility, which rarely occurs in practice.

It is possible to estimate the vector u without taking derivatives, by the method of *finite differences*. Suppose that in the neighborhood of the root we change the parameter q_i by a small amount Δ_i, leading to a change $\Delta_i L$ in the logarithmic likelihood. Then we may estimate the u vector as

$$u = (\Delta_i L / \Delta_i)$$
(A.6.12)

using equation A.6.8 to evaluate the K matrix.

Choice of Δ_i is sometimes delicate, since too large a value may give slow convergence, while too small a value may fail to converge because of rounding errors. Computing programs using equation A.6.12 usually take Δ_i/q_i as a small fraction, say 10^{-4}, subject to change if convergence is not obtained in a small number of iterations.

Gene frequency estimation under panmixia is an example of parameters with a single linear restriction, $\Sigma_{i=1}^{m} q = 1$. We usually impose this by estimating only $m - 1$ gene frequencies, the remaining one (say q_m) being defined in terms of this set as

$$q_m = 1 - \sum_{i=1}^{m-1} q_i.$$

Thus a set of m alleles is of order $m - 1$. The likelihood is arithmetically the same, however q_m is expressed. Let U_i^* be the ML score computed (either exactly or by finite differences) as if there were m independent parameters. From calculus (eq. A.3.1) we know that

$$U_i = U_i^* + U_m^*(\partial q_m/\partial q_i).$$

But $\partial q_m/\partial q_i = -1$. Therefore,

$$U_i = U_i^* - U_m^* \tag{A.6.13}$$

At the root of $U_i = 0$, we see that $U_i^* = U_m^*$, and this is true for all i $(i = 1, \ldots, m)$. Now define

$$E_i = q_i U_i^*$$

as the *expected count* of the ith allele. Summing on both sides,

$$\sum_{i=1}^{m} E_i = \Sigma q_i U_i^* = kN$$

for a sample of size N from a k-somic locus. Since, at the root, U_i^* is a constant, we must have

$$\sum_{i}^{m} q_i U_i^* = U_i^* \Sigma q_i = U_i^*$$

and so from the definition of E_i,

$$q_i = E_i/kN.$$

Writing this as an iteration, the estimate in the nth cycle is

$$q_i^{(n)} = E_i^{(n-1)}/kN, \tag{A.6.14}$$

which does not require calculation of the information matrix K. This is an example of a *counting algorithm*.

We have only scratched the surface of iterative minimization. The most powerful methods replace uK^{-1} in equation A.6.4 by pG^{-1}, where p is a gradient constructed for good convergence and G is an iterative approximation to K. With increasing accessibility of computers, exact differentiation has been replaced by finite differences in all but the simplest cases.

A.7 Tests on Correlations

The general linear hypothesis is

$$Y = \Sigma \, b_i X_i + e, \tag{A.7.1}$$

where Y is a dependent variable, the X_i are independent variables, and e is an error assumed to be normally distributed with mean 0 and variance s^2. The b_i are called *partial regression coefficients*. A test of the hypothesis that $b_j = 0$ can be made on the F ratio of equation A.4.8, where s_1^2 is the error variance when b_j is estimated, and s_2^2 is the value when b_j is set to 0.

The correlation between two variables X_i, X_j is

$$r_{ij} = s_{ij}/s_i s_j, \tag{A.7.2}$$

where the covariance, based on a sample of size N_{ij}, is

$$s_{ij} = \frac{\Sigma(X_i - \mu_i)(X_j - \mu_j)}{N_{ij} - 1}$$

Path analysis provides for each observed correlation r_{ij} an expectation in terms of path coefficients,

$$\rho_{ij} = \sum_k p_{ik} r_{jk}$$

Tests of goodness of fit and estimates of the path coefficients can be made in several ways, as follows. If the samples of N_{ij} pairs are independent, the likelihood in large samples converges to $L = -\chi^2/2$, where

$$\chi^2 = \sum_{i,j}^{m} (r_{ij} - \rho_{ij})^2/\sigma_{ij}^2 \qquad\qquad (A.7.3)$$

and

$$\sigma_{ij}^2 = (1 - r_{ij}^2)^2/N_{ij}.$$

In small samples the approach to normality is better for

$$\chi^2 = \sum_{i,j}^{m} (z_{ij} - \bar{z}_{ij})^2/(N_{ij} - 3) \qquad\qquad (A.7.4)$$

where

$$z_{ij} = \frac{1}{2} \ell n \left(\frac{1 + r_{ij}}{1 - r_{ij}} \right)$$

$$\bar{z}_{ij} = \frac{1}{2} \ell n \left(\frac{1 + \rho_{ij}}{1 - \rho_{ij}} \right)$$

If all the r_{ij} are determined in a single sample of size N, they are not independent. Then the appropriate test is

$$\chi^2 = Z' \Sigma^{-1} Z \qquad\qquad (A.7.5)$$

where Z is the vector of deviations $(z_{ij} - \bar{z}_{ij})$ and Σ is their covariance matrix. In large samples this converges to the Bartlett test,

$$\chi^2 = (N - 1) \left[\ell n \frac{R.}{P.} - m + tr.RP^{-1} \right] \qquad\qquad (A.7.6)$$

where R is the matrix of r_{ij}, P is the matrix of ρ_{ij}, and $tr.RP^{-1}$ is the sum of the diagonal elements of the matrix RP^{-1}.

Minimization of χ^2 gives estimates of path coefficients. Agreement between tests based on r and z increases confidence in the results. These tests are more reliable than ones based on standard errors.

A.8 Sequential Analysis

Linkage and surveillance are commonly analyzed sequentially, assuming a formal alternative to the null hypothesis (fig. A.8.1). Such methods depend on the probability ratio, which we shall consider first for linkage.

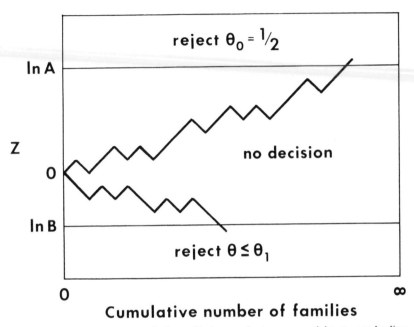

Fig. A.8.1: Sequential tests. A chart of lod scores for two sequential tests, one leading to acceptance and the other to rejection of the null hypothesis.

Let $P(i; \theta_0)$ be the probability of the ith family on the null hypothesis H_0 that $\theta = \theta_0$ (here $\theta_0 = 1/2$), and let $P(i; \theta_1)$ be the probability of the ith family on an alternative hypothesis H_1 that $\theta = \theta_1$ (here $0 \le \theta_1 < 1/2$). Then

$$\lambda_i = \frac{P(i; \theta_1)}{P(i; \theta_0)} \tag{A.8.1}$$

is the *probability ratio* for the ith family. In independent sampling the probability ratio for a set of n families is $\Pi^n_{i=1} \lambda_i$. It is usually easier to add logarithms than to multiply probabilities, and so we define

$$z_i = \log \lambda_i \tag{A.8.2}$$

as the *lod* for the ith family, by elision of *lo*garithm of the o*d*ds. Then

$$Z = \sum_{i=1}^{n} z_i \tag{A.8.3}$$

is the total lod for n families. Published tables of z use common logarithms, $\log_{10}$.

A valuable property of probability ratio tests is that their expected value under H_0 is 1, regardless of H_1:

$$E(\lambda) = \sum_i P(i; \theta_0)\lambda_i \qquad (A.8.4)$$
$$= \sum_i P(i; \theta_1) = 1.$$

Suppose the inequality

$$\lambda > A$$

occurs with probability α under the null hypothesis. Then

$$A\alpha < E(\lambda) = 1,$$

which implies

$$\alpha < 1/A.$$

In terms of lods,

$$P(Z > \log A)|_{H_0} < 1/A. \qquad (A.8.5)$$

Clearly this principle has several advantages, among them its reliability in small as well as large samples, its dependence solely on elementary laws of probability, and the ease with which all kinds of families of known and unknown phase may be combined simply by addition of their lods. In samples of fixed size θ_1 may be replaced by $\hat{\theta}$, the ML estimate, to give a confidence interval.

Because families that give information about a particular pair of systems G and T are usually collected over a period of time by several investigators, it is convenient to have a test that will permit detection or exclusion of linkage (or any specified value of linkage) with the smallest possible number of families. This requires that linkage be tested *sequentially* after each family or batch of families is added. Let A and B be two preassigned positive numbers, with $A > 1$ and $B < 1$. Then if

$$\log B < Z < \log A, \qquad (A.8.6)$$

the data are indecisive at the chosen significance levels, and more families must be collected to reach a decision. If after the addition of another family or batch of families Z exceeds log A,

$$Z > \log A$$

this evidence for linkage is significant. If however Z becomes less than log B with additional data,

$$Z < \log B,$$

then the evidence is significant that the recombination fraction exceeds θ_1, no test having been made of loose linkage ($\theta_1 < \theta < 1/2$). Thus the investigator may test with great efficiency for close linkage (say, $\theta_1 < .1$) when a small sample is expected or for loose linkage ($\theta_1 > .1$) in a large sample.

By equation A.8.5, we are entitled to state that

$$P(Z > \log A)|_{H_0} < 1/A$$

and

$$P(Z < \log B)|_{H_1} < B.$$

The theory of sequential tests gives a more precise result. Suppose the amount of information in an increment of data is so small that H_0 is accepted at log B or rejected at log A, the excess over the boundaries being negligible. If α is the probability of rejecting H_0 when it is true (a *type I error*) and β is the probability of rejecting H_1 when it is true (a *type II error*), then at the first boundary

$$\lambda = \frac{1 - \beta}{\alpha} = A, \quad \text{and so} \quad \alpha = \frac{1 - \beta}{A}. \tag{A.8.7}$$

At the second boundary

$$\lambda = \frac{\beta}{1 - \alpha} = B, \quad \text{and so} \quad \beta = B(1 - \alpha).$$

Even when the excess over the boundaries is not negligible, Wald has shown that the conditions (eq. A.8.7) hold to a close approximation if α and β are small, as is always the case in practice. Thus the significance levels also apply to batches of several families and even to nonsequential samples.

The power function $P(\theta)$ is the probability of rejecting H_0 as a function of the true value of θ, neglecting the excess over the boundaries. Clearly

$$P(\theta_0) = \alpha \tag{A.8.8}$$

$$P(\theta_1) = 1 - \beta$$

For other values of θ, Wald provided the simultaneous equations

$$\sum_i P(i; \theta)\lambda_i^h = 1 \tag{A.8.9.}$$

$$P(\theta) = \frac{1 - B^h}{A^h - B^h}. \tag{A.8.10}$$

We first find the value of h that satisfies equation A.8.9 for a given value of θ, and then substitute h in equation A.8.10.

Wald also showed how to calculate the average sample size (here, number of families) to complete a sequential test. For any value of θ, the probability of reaching the boundary log A is $P(\theta)$, and so the expected sample size is

$$E_\theta(n) = \frac{P(\theta)\log A + [1 - P(\theta)]\log B}{E_\theta(z)} \tag{A.8.11}$$

where the expected lod is

$$E_0(z) = \sum_i P(i; \theta)z_i$$

and

$$z_i = \log \frac{P(i; \theta_1)}{P(i; \theta_0)}.$$

In particular,

$$E_{\theta_1}(n) = \frac{(1 - \beta)\log A + \beta \log \beta}{E_{\theta_1}(z)}$$

and

$$E_{\theta_0}(n) = \frac{\alpha \log A + (1 - \alpha)\log B}{E_{\theta_0}(z)}.$$

The two functions $P(\theta)$ and $E(n)$ determine the best sequential test for a particular purpose and the extent of its superiority over nonsequential procedures. Requirements to impose on these functions are suggested by the probability distribution of θ and/or the consequences of type I and type II errors.

We want a significant linkage to be reliable, given that only about .054 of random loci are on the same chromosome. It is not critical to exclude linkage, since routine collection of data does not stop even when linkage is "rejected." Therefore it is customary to take $A = 1000$, $B = .1$, or equivalently $\log A = 3$, $\log B = -1$. Then we say that linkage is significant when $Z > 3$, and that linkage is reasonably excluded (in the absence of other evidence) when $Z < -1$, being indecisive so long as $-1 < Z < 3$.

For surveillance, we know less about the distribution of the risk parameter. We are anxious to detect a hazard when it occurs, but to avoid false alarms. It is therefore customary to choose a test that will lead to a decision about once every two years under the null hypothesis of constant risk, with $A = B = .1$, so that a false alarm will occur about once per 20 years. An alarm leads first to supplementary tests. If the hazard is confirmed, intervention to reduce risk will ultimately be attempted.

Subject Index